Felix Publishing 2017
www.felixpublishing.com.au
email: info@felixpublishing.com
Print copies available from publisher.

Fossils - Life in The Rocks
Part of the Series **Adventures in Earth Science**
Other books in the series include:
> Exploration Science (Field Geology and Mapping)
> Riches from the Earth (Minerals, Mining & Energy)
> Changing the Surface (Erosion and Landscapes)
> Rocks - Building the Earth
> A Dangerous Planet (Earth Hazards)
> Through Sea and Sky (Oceanography and Meteorology)
> Beyond Planet Earth (Astronomy)

2016 digital book release
ISBN: 978-0-9946433-3-9
Print Edition
ISBN: 978-09946432-4-7
Author: Dr P.T.Scott

All illustrations, photographs and videos by the author unless stated.
Cover photo: Insect embedded in amber. Design based upon that of AJS
Creative, Brisbane.

Registration:
Thorpe-Bowker +61 3 8517 8342
email: bowkerlink@thorpe.com.au

FELIX
PUBLISHING

FOSSILS - LIFE IN THE ROCKS

Dr. Peter T. Scott

First released 2017

To my grandchildren who are
yet to find their own adventures.

About the Author

Dr. Peter Scott is an award-winning teacher of Earth Science of over forty years' experience in both Secondary and Tertiary Education. He holds a Bachelors' Degree, two Masters' Degrees and a Doctorate including many years on his own research in locating and correlating coal measures. He has visited many places of interest including Antarctica, the Andes, the Amazon, North Africa, volcanic islands of the Pacific and Asia, California northern Europe and remote Australia.

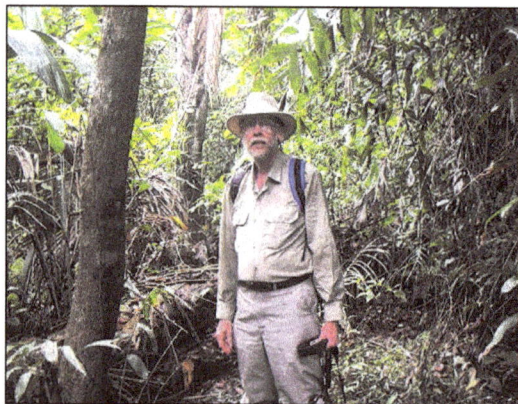

Dr. Scott, in the jungle near Puerto Maldonado, Peruvian Amazonia, Peru, 2011

Table of Contents

Chapter 1:   The Formation of Fossils

1.1 Introduction

Some **sedimentary rocks** were found to contain small structures which did not look like they had been formed from sedimentation or post-sedimentary processes. In fact, some looked like grotesque lifeforms, and were considered the work of the devil to many who accidentally came upon them. Those with more curiosity and powers of observation saw these objects for what they were - the remains of past living creatures that had somehow became embedded in the rock. The ancient Greek philosophers **Xenophanes** (570-480 BC) and **Aristotle** (384-322 BC), thought that fossil sea shells found on land had once been under the sea, and the Persian Naturalist **Ibn Sina** (also known as **Avicenna**: 980 – 1037 AD), wrote about many aspects of geology, including fossils. The artist, inventor, anatomist and geologist, **Leonardo da Vinci** (Italian:1452-1519 AD), also believed that fossil shells and fish were the remains of living things, and that the age of the Earth was much greater than what was believed at that time.

In Europe, the systematic study of **fossils** began with the overall interest in natural philosophy which occurred during the end of the 18th century. **Georges Cuvier** (French: 1769-1832) noted the similarities and differences between the bodies of these ancient lifeforms to those of modern living thing, and established the concept of **comparative anatomy** as a scientific discipline. In showing that some fossil animals did not resembled any modern living species, Cuvier suggested that ancient plants and animals could become **extinct**. This led to the development of the new study called **palaeontology,** from Greek, *palaios* for ancient, *ontos* for creature, and *logos* for study, a name coined in 1822 by **Henri Marie Ducrotay de Blainville** (French: 1777 - 1850), to refer to the study of ancient living organisms.

It was soon discovered that knowledge of fossils could be useful in the new industrialisation which was sweeping Europe in the 19th Century. This was especially true in Great Britain, where the collection and study of fossils not only became a popular pastime, but also a practical aid in finding coal and for comparing different rocks involved in the engineering of building canals and bridges.

1.2 Preserving Life

Almost any remainder of past life preserved and found in modern times can be classed as a fossil, a name from the Latin, *fossilis* meaning to obtain by digging. The predominant remains found are those of the hard parts of dead organisms, such as shells, bones, sponge spicules, and teeth. Soft

Figure 1.1 A reconstruction of an ancient Mastodon trapped in a tar pit

parts such as plant tissue, flesh, feathers and scales, as well as impressions made by footprints or burrowing, must be covered quickly or changed to some resistant form before the processes of decay or disturbance begins. There have been a few rare instances where the soft parts of animals have been preserved, such as the flesh of mammoths frozen in the frozen soil of the permafrost, insects in amber which is a fossilized tree sap, and a few mammal specimens from tar pits. Most of the fossils discovered have undergone some change which has enabled them to be preserved. The most common changes are:

- **Permineralization** is due to the crystallization of minerals from groundwater which have been absorbed by the porous specimen, filling in the spaces within it. The new mineral is often the same as that found in the original animal's shell or bone, and is deposited in the same orientation as that of the original mineral.

Figure 1.2: A small (4 cm) ammonite permineralized with silica

- **Replacement** occurs when ground water solutions dissolve out the mineral of the organism and replaces it, crystal by crystal, by some other, more resistant mineral.

Figure 1.3: Fossilized plants replaced with iron oxides

- **Recrystallization** occurs when a chemical change occurs within the mineral of the fossil, converting it to a more stable form e.g. in many shelled organisms, aragonite the original form of calcium carbonate, is converted to the more stable and resistant calcite.

Figure 1.4: The aragonite which formed the theca (hexagonal walls) of the living coral has been converted to calcite and filled in

- **Carbonization** is common with plant fossils when the plant body decomposes to a thin film of black carbon within the sediment.

Figure 1.5: A specimen of the plant *Pecopteris* showing carbonized fronds (x ½)

- **Impressions** are the remains of the internal and/or external parts of the organism as an impression after the organism itself has decayed. The impression or hollow, called a **mold,** may be found by itself or filled by material which then forms a replica, or **cast**, of the original.

Figure 1.6: The mollusc *Keenia* as a mold (left) and as a cast (right)

- **Encasement** occurs when the organism is trapped within a preserving substance e.g. mammoths in the permafrost or frozen soil of Russia, in the La Brea tar pits in America, or insects in amber which is ancient tree sap, from the Baltic region of Europe.

Figure 1.7: Fossils preserved in polished amber

- **Ichnofossils** are named from the Greek, *ikhnos*, for trace or track and are also called trace fossils. These include tracks, burrows, **coprolites** (solid waste pellets) and eggs, which may be later mineralized or form as molds and casts.

Figure 1.8 Worm burrows preserved as traces in a fine marine mudstone

Figure 1.9: A footprint 5 cm. long of a small Theropod (dinosaur) from Winton, Queensland, Australia

1.3 Pseudofossils

These are structures found within rock which resemble fossils but are not the remains of ancient lifeforms. They are structures caused by weathering or mineralization and are termed pseudofossils or false fossils. Some of the most common pseudofossils are:

- **Dendrites** which look like moss fossils but are really tiny fracture patterns filled with black manganese dioxide as the mineral pyrolusite.

  Figure 1.10: Fine dendrites in a limestone

- **Concretions** occur in sedimentary rocks such as sandstones, and form early in the sedimentation process by precipitation of secondary minerals, often iron oxides such as haematite or siderite, within the porous rock. These form in concentric layers around a nucleus within the rock and then are exposed as the softer rock weathers away.

Figure 1.11: An iron-stone concretion

- **Cone-in-cone structures** are composed of concentric cones inserted inside each other, mostly composed of calcite or gypsum mineral, with thin layers of clay between the cones. They are possibly caused by crystal fibre growth and displacement, as the sedimentary rock is deeply buried or undergoes pressure. There is still some doubt as to their formation.

Figure 1.12: Cone-in-cone structures

- **Fulgurites** from the Latin, *fulgur*, meaning lightning, are found at the base of pure silica sand dunes. Once thought to have been formed by the borrowings of worms, they are actually the result of lightning strikes on the top of the dune which fuses the silica sand together.

Figure 1.13: A small fulgurite about 10 cm long

1.4 Collecting Fossils in the Field

Fossils are not always easy to find. This is because of the:

- lack of lifeforms in the area originally

- lack of sedimentation to bury the lifeform and preserve it

- decomposition of the lifeform under normal conditions of decay

- disturbance at a later period due to volcanism, metamorphism, Earth movement and erosion which destroys the **fossil horizon** - the thin layer which contains the fossils.

Remember that fossils are:

- Difficult to find because the fossils may be located only within a very thin fossil horizon of a few millimetres or centimetres thickness. This horizon may be within metres of unproductive layers of sedimentary rock, and so the horizon itself is difficult to locate.

- Delicate because of their method of preservation and/or the rock type in which they are contained. This is especially true if the containment rock is a fine mudstone or shale. In that case, only a minimum of rock immediately around the margins of the fossil can be chiseled out, and then the entire specimen is carefully removed, wrapped and packed. Larger fossils are often coated with a strengthening compound, such as layers of plaster with reinforcing material, before being carefully cut out of the rock and packed into boxes.

- Scarce, so in the field only the minimum number of specimens should be collected with minimal damage to the environment. Digging within a fossil horizon should be done very carefully so as to minimize the risk of destroying other fossils. Once one specimen is found, the fossil horizon or layer containing the fossils can usually be determined. Many fossil sites are protected or are on private property, so one should not be collecting in these areas without permission.

- Used for study and/or display, so once collected, fossils should be identified against known databases or prepared for display or storage. This may be a simply a matter of reference to some good texts, or it might mean professional assistance with a comparison with an existing collection at a university or museum.

◼️◀ Online Video 1.1: Collecting plant fossils. Go to
https://youtu.be/OHlmx3iBhf0

Chapter 2: Classifying and Naming Fossils

2.1 Introduction

Fossils are given names in the same way as are modern organisms using the **binomial system** meaning two names, which was first developed by **Linnaeus (Karl von Linne**, Swedish: 1707 - 1778). In this system, organisms are classified into major groups such as empire, kingdom, phylum (or division for plants), class, order, family, genus and species. An individual organism is identified by their genus and species, hence the term binomial system. There may also be further sub-divisions of these **taxa**, a name derived from the **taxonomy,** the classification of organisms, and scientists use variations of this system depending upon their field of study. As well as naming a new fossil, its discoverer would also describe its anatomical details, possible geological age and the **type locality** where it can be found.

2.2 Classification of Living Things

A full, global classification on all life is beyond the scope of this book, as there are representatives of most classification divisions within the fossil record. Systems of taxonomy change as new information is found. The classification system given here is a simple generalization, with an attempt to use the most modern taxonomies applied to the fossil record. It does, however, include the additional expansion of older systems, to take into account more modern studies of single-celled organisms which are also found in the fossil records. This **Six Kingdom Model** is based on the work of **Thomas Cavalier-Smith** (English: 1942 - see Cavalier-Smith, T., 2007. A revised six-kingdom system of life. *Biological Reviews* 73 (3): 203-66.) and modified here for simplicity. It begins with a division of life into two major groups called Empires:

- Empire Prokaryote which are organisms having no **nuclear membrane.** They include bacteria, cyanobacteria (blue-green algae) and the archaeobacteria (usually only found in water with extremely high temperatures and acidity).

- Empire Eukarya which are organisms which have cells with a nucleus and organelles, or structures within the cell, which are contained in membranes.

These empires are then sub-divided into a separate kingdom for bacteria and then the main kingdoms of other living things. A simplified view of the major taxonomic groups for ease of description could be given as:

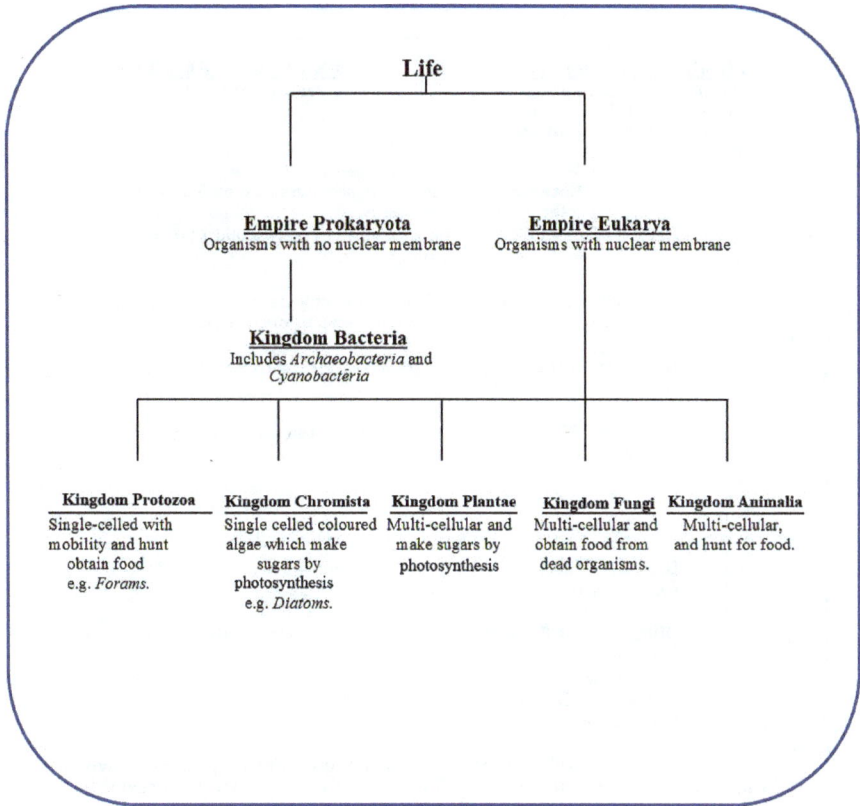

Figure 2.1: A simplified classification of living things

Continuing this classification further for plants, or Kingdom Plantae:

KINGDOM PLANTAE

SUB-KINGDOM BRYOPHYTA
Do not have vascular tissue
(special tubes to carry air,
Water & food) e.g. *mosses*
Liverworts

SUB-KINGDOM TRACHEOPHYTA
Have vascular tissue

Note the first 8 Divisions are classified as Gymnosperms - plants with naked seeds.

— **DIVISION PSILOPHYTA** - most primitive of vascular plants & extinct. Branching leafless stems & no true roots.

— **DIVISION LYCOPODIOPHYTA** - branching stems with numerous small leaves & spores produced by a club-shaped organ e.g. *club mosses*.

— **DIVISION SPHENOPHYTA** - jointed stems with whorls of leaves e.g. Horsetails such as *Equisetum*.

— **DIVISION PTERIDOPHYTA** - feathery leaves as fronds with spore capsules on their underside e.g. Ferns

— **DIVISION CYCADOPHYTA** - palmlike and have seeds in cones e.g. Cycads.

— **DIVISION GINKGOPHYTA** - fan-shaped leaves with seeds in fleshy, rounded structures e.g. *Ginkgo*.

— **DIVISION GNETOPHYTA** - strap-like leaves with simple vascular tissue and seeds in cones e.g. *Welwitschia*.

— **DIVISION CONIFEROPHYTA (Pinophyta)** - evergreens with needle-like leaves and seeds in cones e.g. Conifers.

— **DIVISION ANGIOSPERMOPHYTA (Magnoliophyta)** - largest Division with covered seeds in flowers e.g. CLASS MONOCOTYLEDONAE (one cover) long leaves with parallel veins and fibrous roots e.g. grasses and cereals; CLASS DICOTYLEDONAE (two covers) with broader leaves with network veins and large tap root systems e.g. most common large flowering bushes and trees such as Eucalyptus, Beech, roses etc.

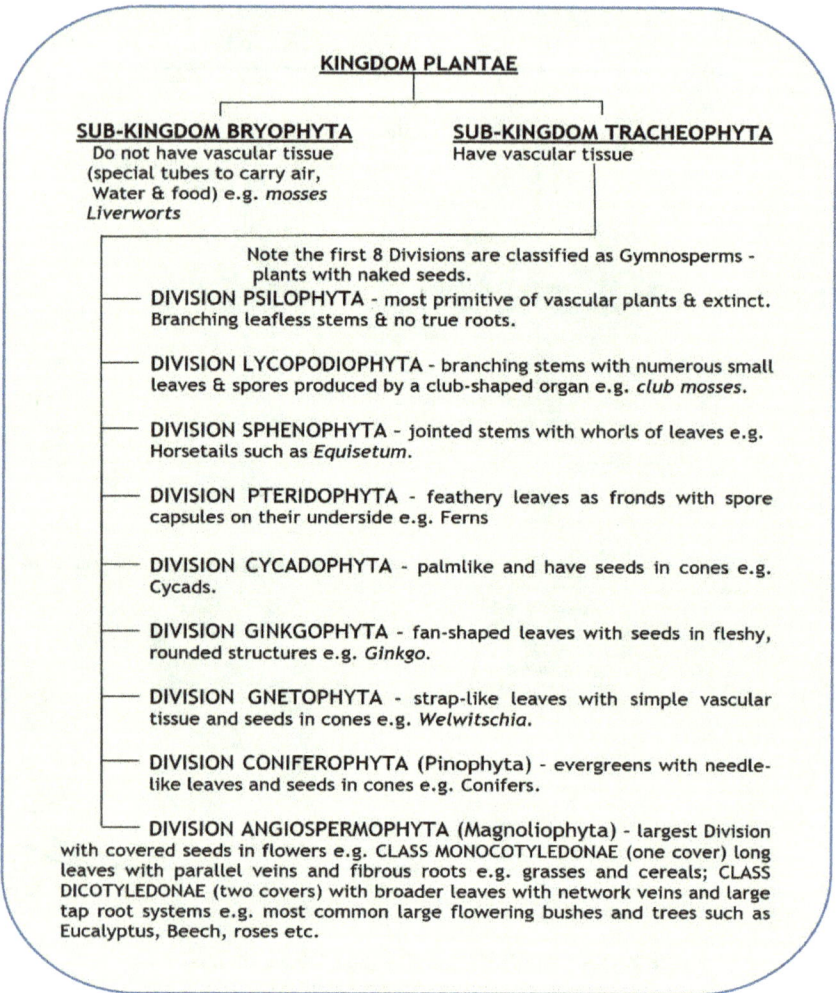

Figure 2.2 A simplified classification of plants

Continuing for animal classification:

KINGDOM ANIMALIA

SUBKINGDOM RADIATA
Bodies have radial symmetry

SUBKINGDOM MYXOZOA
Aquatic parasites

SUBKINGDOM MESOZOA
Marine worm-like parasites

SUBKINGDOM BILATERIA
Bilateral symmetry - alike on both sides of a central line

PHYLUM PORIFERA
Undifferentiated cells; hollow body with pores on outer surface e.g. sponges

PHYLUM CNIDARIA
Marine, hollow body, stinging cells e.g. jellyfish, corals

PHYLUM CTENOPHORA
Body cavity & move using cilia e.g. comb jellyfish

SUPERPHYLUM DEUTEROSTOMA
Embryos develop their anus first

SUPERPHYLUM ECDYSOZOA
Three-layered outer shell which grows by moulting

SUPERPHYLUM PLATYZOA
Flat, embryos develop mouth first (Protostoma)

SUPERPHYLUM LOPHOTROCHOZOA
Protostoma with hair-like cilia around mouth

PHYLUM HEMICHORDATA
Marine, colonial & have a Simple nerve chord

PHYLUM ECHINODERMATA
Spiny outer skin with single hole For excretion and feeding

PHYLUM CHORDATA
Have a dorsal spinal chord

PHYLUM NEMATODA
Cylindrical bodies with internal organs. Often Parasitic e.g. round worms

PHYLUM ARTHROPODA
Exoskeleton with jointed limbs e.g. crustaceans, spiders, insects & CLASS TRILOBITA

PHYLUM PLATYHELMINTHES
No specialized respiratory nor circulatory system. e.g. flatworms

PHYLUM BRYOZOA
Colonial marine filter feeders e.g. Fenestella

PHYLUM ANNELIDA
Segmented cylindrical bodies e.g. leeches & earthworms

PHYLUM BRACHIOPODA
Hinged shells with extruding Foot from a hole in one shell

PHYLUM MOLLUSCA
Have a mantle of tissue e.g. clams, squid etc.

SUBPHYLUM TUNICATA
Marine filter feeders Attached to the sea floor e.g. Sea Squirts

SUBPHYLUM CEPHALOCHORDATA
Segmented with long bodies e.g. Amphioxus

SUBPHYLUM VERTEBRATA
Nervous chord within bony

SUPERCLASS AGNATHA
Lacking jaws e.g. Hagfish, Lampreys

SUPERCLASS GNATHOSTOMATA
Have jaws

CLASS PLACODERMI
Extinct armoured fish

CLASS CHONDRICHTHYES
Cartilaginous fish e.g. sharks & rays

CLASS OSTEICHTHYES
Bony fish e.g. salmon, tuna etc.

CLASS AMPHIBIA
Lay eggs in water but move to land e.g. frogs, toads

CLASS REPTILIA
Dry, scaly skin e.g. snakes, lizards, alligators

CLASS AVES
Light bones e.g. birds

CLASS MAMMALIA
Suckle young on milk e.g. humans whales, bats etc.

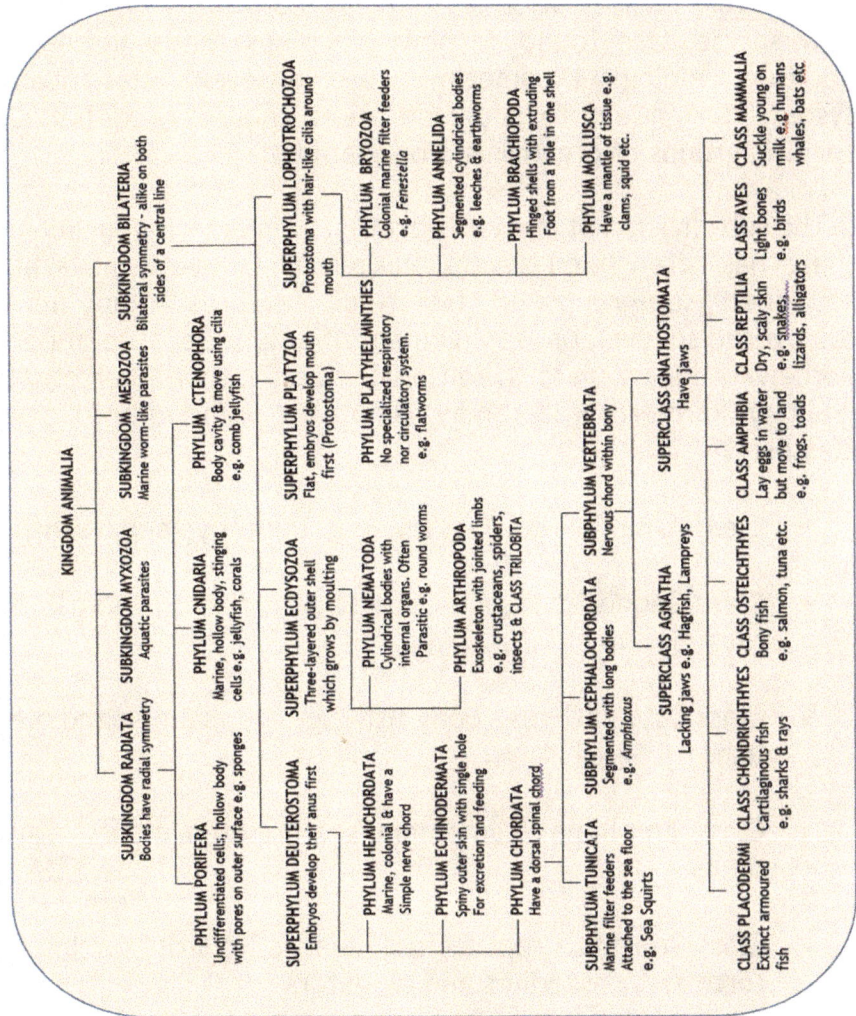

Figure 2.3: A simplified classification of animals

<u>Remember</u>: Classification systems are constantly changing; as new species are found or new groupings are suggested. In addition, biologists sometimes use different classification systems as an aid to show different relationships between specific groups of organisms. For example:

In the sub-division of the **vertebrates**, animals with backbones, there has often been several variations as taxonomists have redefined sub-groups in terms of features from newly-discovered species, or reassignment of groups because of newly perceived ideas. In a traditional classification, vertebrates (Subphylum Vertebrata) may simply be divided into seven classes:

- Class Agnatha - jawless fish e.g. lampreys and hagfish

- Class Chondrichthyes - cartilaginous fish e.g. sharks and rays

- Class Osteichthyes - bony fish e.g. cod, salmon and other fish

- Class Amphibia - amphibians e.g. salamanders, frogs, toads

- Class Reptilia - reptiles e.g. snakes, lizards, turtles and tortoises, crocodiles and alligators

- Class Aves - birds e.g. doves, eagles and other birds

- Class Mammalia - mammals e.g. humankind, apes, monkeys, bats, whales/dolphins and others.

Recent research into the nature of dinosaurs, as another example, has shown that some of these animals were more like mammals, and some probably evolved into modern birds. Some classifications use different classes to include dinosaurs:

- Class Sauropsida - reptiles and birds, as well as the larger dinosaurs such as *Brachiosaurus and Diplodocus*

- Class Synapsida - the mammal-like dinosaurs e.g. *Dimetrodon*

When the Linnaean System is used, it is usual to give only the genus and then the species to identify the organism, with the species written in lower case and both names in italics (if printed) or underlined (if hand-written). For example, for humankind the name would be:

Homo sapiens

The full taxonomy for modern humankind could be given as:

DOMAIN: Eukarya - having cellular structure in which cells have cell nuclei encased in a cell membrane

KINGDOM: Animalia - multicellular organism which obtain food from other organisms

SUBKINGDOM: Metazoa - multicellular

SUPER PHYLUM: Eumetazoa - true tissues and organs

DIVISION: Vertebrata - bony vertebral column

PHYLUM: Chordata - spinal cord of nerve tissue

SUBPHYLUM: Vertebrata - has a bony vertebral column

CLASS: Mammalia - warm-blooded, suckles young on milk

ORDER: Primata - forward-facing eyes and grasping fingers

FAMILY: Hominidae - upright with a large brain

GENUS: *Homo* - curved spine and human features

SPECIES: *sapiens* - having the ability to think

SUB-SPECIES: *sapiens* – reinforcement of the ability to think

A **species** is defined as that group of organisms in which the individuals may mate and give rise to new offspring. In some texts, the term is abbreviated to sp. when referring to an organism with an uncertain classification (e.g. *Dicroidium sp.*). It also may be necessary to further sub-divide the species into a sub-species or variety in order to distinguish between two

Figure 2.4: Specimen of *Dicroidium odontopteroides*, variety odontopteroides (Morris)

slightly different organisms. Sometimes the name of the discoverer is also attached to the name of the organism. For example, the seed-fern, *Dicroidium odontopteroides* has several varieties including *D. odontopteroides (Morris)*.

Chapter 3: Geological Time Scale

3.1 Introduction

As interest in fossils developed in the early 19th Century, paleontologists in Britain found that they could compare similar rocks from one place to another using the groups of fossils found within them. This concept was first suggested by **William Smith** (English: 1769-1839 who was called "Strata Smith" by his colleagues), the father of British Geology, and further developed by **Sir Charles Lyell** (Scottish: 1797-1875). A relative age comparison for fossils could also be determined because of the **Law of Superposition,** devised by **Nicholas Steno** (1638-1686 a Danish scientist and priest), which stated that in a vertical sequence of strata, the older rocks were put down first and then these were covered on top by younger layers, the youngest being on the top.

In the diagram below, a general sequence for the age of rocks in these localities could be determined using the different fossils contained within them. Each fossil image represents a group of fossils or an **assemblage,** commonly found together. The rock stratigraphic columns may not necessarily be the same for each location, but they contain some rock layers common to several locations. There are also some layers without fossils, representing a time when sedimentation did not trap any lifeforms, and so some fossils are missing from some locations. This is probably because there was a time gap from one period of sedimentation to another. Such a break or hiatus in sedimentation is called an **unconformity**. The same fossil assemblage can sometimes be matched from one location to another giving these strata the same relative age. This technique is called **correlation**. In the 19th century, when this concept was being developed, palaeontologists assigned time names for the fossil assemblages rather than numbers to represent the relative ages of various fossil assemblages. At this time, there was no way of finding out the true, chronological age for any rock or fossil. Between 1820 and 1850, the geological time **periods** known today, were described and named by a number of British and European palaeontologists. Where rock units allowed a further sub-division in age, the terms Early, Middle and Late were also applied to these Periods. For example, the plant species *Dicroidium* is known from the Late Triassic Period.

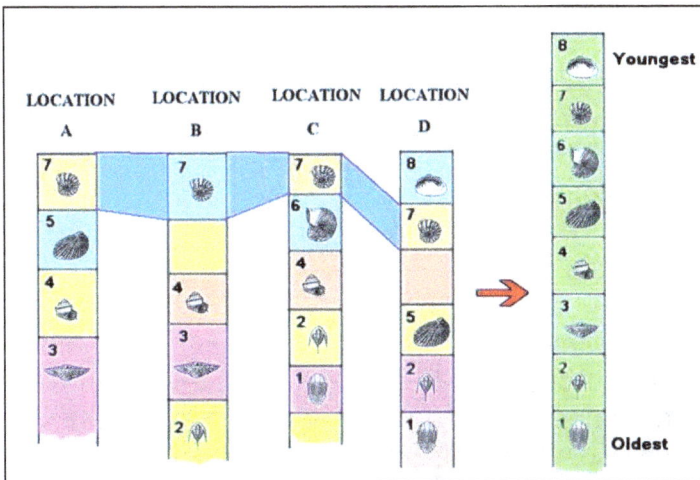

Figure 3.1: Diagram showing how fossils can be compared at different locations

3.2 A Modern Approach

After the discovery of radioactivity in 1898 by **Antoine-Henri Becquerel** (French: 1852-1908), the actual age or absolute time, could be estimated for each Period. Radioactive age determination makes use of the theory of radioactive half-life of particular unstable elements found within rock-forming minerals.

The **half-life** of an element is the time required for a known amount of radioactivity or mass of the element to decay to half of its original radioactivity or mass.

As the radioactive elements decays, they produce a series of unstable elements and finally decay to stable elements such as lead. If the ratio of radioactive element to stable products can be determined in a mineral sample, then the absolute age of the mineral, and thus the rock in which it was found, can be determined. There

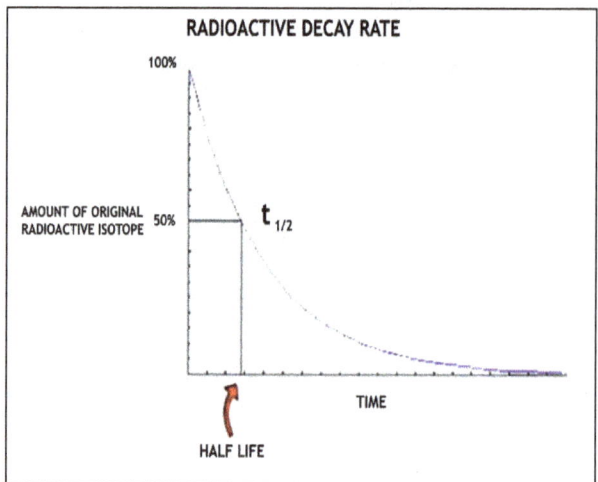

RADIOACTIVE DECAY RATE

100%

AMOUNT OF ORIGINAL RADIOACTIVE ISOTOPE 50% $t_{1/2}$

TIME

HALF LIFE

Figure 3.2: Graph showing the decay of a radioactive isotope and how the half-life ($t\frac{1}{2}$) is found. For isotopes within rocks, the decay time is very large, so real time measurements are taken and the graph time axis is extended

is, however, the possibility of errors due to loss or gain of any of the original radioactive material or its products during the analysis.

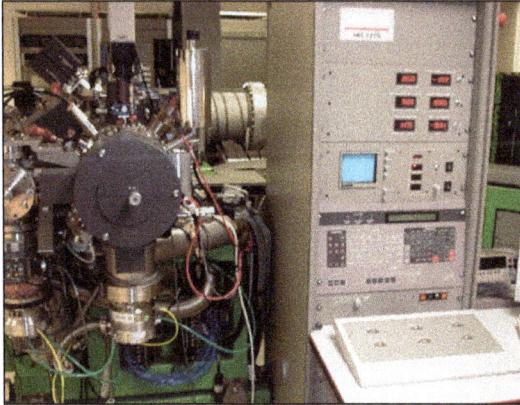

Figure 3.3: Mass spectrometer used for determining radiometric age at the Geology Faculty, University of Edinburgh, Scotland

The most commonly used radioactive dating methods are:

RADIOACTIVE ISOTOPES	HALF-LIFE	DATING RANGE	MATERIALS THAT CAN BE DATED
Uranium 238 to Lead 206	4.50×10^9 years	10^7 to 10^9 years	Zircons in Igneous Rocks and Uranium Ores
Uranium 235 to Lead 207	0.71×10^9 years		
Potassium 40 to Argon 40	1.30×10^9 years	10^4 to 10^9 years	Micas, Hornblende, Volcanics, Sandstone and Siltstone
Rubidium 87 to Strontium 87	4.7×10^{10} years	10^7 to 10^9 Years	Micas and Metamorphic Rocks
Carbon 14 to Nitrogen 14	5730 years with an error of + or - 30 years	0 to 5×10^4 years	Wood Tissue, Bone, Shells and Water

Table 3.1: Some of the common radiometric dating methods

By the middle of the 20th Century, a modern geological time scale with absolute ages could be constructed:

EON	ERA	PERIOD		EPOCH
PHANEROZOIC EON (from Ancient Greek *fanerós* and *zoi*, meaning *visible life*)	CENOZOIC ERA (from *Kainos* meaning *Recent Life*)	QUATERNARY (Fourth major group of Periods) 2.58 million years ago until present		HOLOCENE or Recent (*Holos - Complete*) 0 to 0.0117 m.y.a.
				PLEISTOCENE (*Pleiston-Most*) refers to amount of shells to 2.58 m.y.a
		TERTIARY (Third major group of Periods following after the two older groups below) 2.58 to 65.5 m.y.a.	NEOGENE (*New Born*) 2.58 to 23.03 m.y.a.	PLIOCENE (*Pleion-More*) 5.33 m.y.a
				MIOCENE (*Meion - Less*) 23.03 m.y.a
			PALEOGENE (*Ancient Born*) 23.03 to 65.5 m.y.a.	OLIGOCENE (*Oligos - Few*) 33.9 m.y.a
				EOCENE (*Eos - Dawn*) 55.8 m.y.a
				PALAEOCENE (*Palaeo - Old*) 65.5 m.y.a.
	MESOZOIC (from *Meso* meaning *Middle Life*)	CRETACEOUS PERIOD (From Latin *Creta* - Chalk) 65.5 to 145.5 m.y.a.		EPOCHS too numerous to detail here
		JURASSIC PERIOD (from the Jura Mountains, France) 145.5 to 199.6 m.y.a.		
		TRIASSIC PERIOD (from the three rock layers of Germany) 199.6 to 251.0 m.y.a.		
	PALAEOZOIC (from *Palaeo* meaning *Old Life*)	PERMIAN (from the Perm region of Russia) 251.0 to 299.0 m.y.a.		EPOCHS too numerous to detail here
		CARBONIFEROUS (Coal-bearing) In America, this is divided into two Periods there: Pennsylvanian 299.0 to 318.1 m.y.a. & Mississippian 318.1 to 359.2 m.y.a.		
		DEVONIAN (After Devon, England) 359.2 to 416.0 m.y.a.		
		SILURIAN (Celtic Tribe - Silures) 416.0 to 443.7 m.y.a.		
		ORDOVICIAN (Celtic Tribe - Ordovices) 443.7 to 488.3 m.y.a.		
		CAMBRIAN (Ancient name for Wales) 488.3 to 542.0 m.y.a.		
CRYPTOZOIC EON (*Hidden Life*)	Little is known about this Eon even though it comprises 80% of known time. It is sometimes divided into several Eras: PROTEROZOIC (Earlier Life - 2500-542 years before the Modern Era) ARCHEOZOIC (Ancient Life - 4000-2500 Ma) and HADIAN (Hades - the underworld - hot, no life - 3950-4500 Ma). This era is generally known as the PRECAMBRIAN ERA			

Table 3.2: The modern geological time scale

An analysis of the fossils found in rocks of various ages, suggested that there were certain dominant lifeforms for some of the geological periods. Palaeontologists were able to identify certain specific fossils, called **index fossils**, which were representative of a particular geological period. The best index fossils are those which are commonly found, easy-to-identify at the species level, have a narrow age range, and have a broad geographical distribution. Some of the major groups of fossils which dominated certain geological time periods were:

Cambrian	Trilobites
Ordovician	Graptolites
Silurian	Corals
Devonian	Fish
Carboniferous	Brachiopods
Permian	Early Plants
Triassic-Jurassic-Cretaceous	Dinosaurs
Tertiary	Foraminifera
Quaternary	Mammals

The oldest evidence for life may be the 3.5-billion-year-old sedimentary structures from Australia which resemble modern-day **stromatolites**. Today, these are living mats of microorganisms, mostly **cyanobacteria** or the blue-green algae, which trap thin layers of sediment with their sticky filaments and grow upward to get light for **photosynthesis**. Modern-day examples of stromatolites can still be found in the shallow waters in Shark Bay, Western Australia.

At Jenolan Caves, New South Wales, Australia, visitors can view a rare formation of stromatolites which have formed in layers of calcite crystal in the wet cave environment. Within the Nettle Cave, several large colonies are clearly visible - their odd shape earning them the nick-name of lobster tails. There is some evidence that the stromatolites at Jenolan Caves may be between 20,000 and 100,000 years old.

Figure 3.4: Present-day Stromatolite colonies in Shark Bay, Western Australia

Figure 3.5: Stromatolite structures in the Nettle Cave, Jenolan Caves, New South Wales, Australia

Chapter 4: Plant Fossils

4.1 Introduction

Fossils are often not easy to find unless one knows about the local fossil beds. Even in these localities, it is a matter of geological skill and some luck to find a good specimen, and palaeontologists usually spend years in the one location looking for the thin **fossil**

Figure 4.1: Thin fossil horizon on top of bedding (shown) at Ipswich, Queensland, Australia.

horizon which may contain fossils. Plant fossils are perhaps the easiest of fossils to find but perseverance is still needed as the fossils are usually found in very thin bands within banded sedimentary rocks. Many sedimentary beds are totally unproductive and yield no fossil remains at all.

4.2 Some Common Plant Fossils

Many of the early plant fossils found today were formed before the time of the true, flowering plants. These ancient plants belong to a world of giant ferns, cycads, palms and conifers, and although many structures look like modern leaves and stems, most of these ancient plants have only simple structures with which to transport water, food and air and to reproduce. What might look like a modern stem may be a simple support column called a **rachis.** A simple structure resembling a collection of leaves, like a fern frond, are called **pinnae** which consist of smaller, individual leaf-like structures called **pinnules.** Most of the primitive plants, such as ferns, reproduced using **spores** rather than true seeds and do not have the complex **vascular tissue** found in more advanced plants. Some of the most common of fossil plants include:

- *Lepidodendron* species named from the Ancient Greek, *lipido* and *dendron,* for scale tree, known from the Carboniferous Period and belonging to Division Lycopodiophyta. It reached heights of over 30 metres, and trunks were covered with **leaf scars** which were possibly green. At the top, the trunk would have split into smaller branches with long, narrow leaves.

Figure 4.2: *Lepidodendron* fossil of trunk leaf scars and a reconstruction of the plant

- *Glossopteris* species named from Ancient Greek, *glossa*, meaning tongue, from the Permian Period of the Southern Hemisphere. It was a seed-fern from Division Pterophyta, with distinctive long, extended oval or **spathulate** fronds, with veins branching out then re-joining i.e. having **reticulate venation**. Another fossil, *Vertebraria*, was found to be the roots of *Glossopteris* species.

Figure 4.3: *Glossopteris* leaf and reconstruction of the plant

- **Gangamopteris** was a genus of Carboniferous-Permian plants, very similar to Glossopteris but lacking a strong midrib. Gangamopteris dominates many of the Permian and Carboniferous coal deposits of the world.

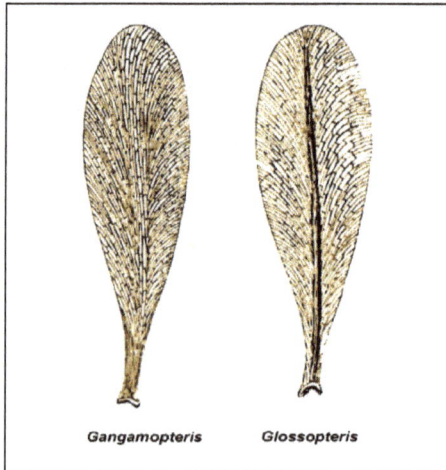

Figure 4.4: *Gangamopteris* (left) compared to *Glossopteris* (right)

- *Pecopteris* species named from Ancient Greek, *pekin,* meaning to comb and *pteris* meaning a fern, known from the Devonian Period and belonging to Division Pterophyta of ferns. It had branches with multiple pinnules alternating on two sides of the rachis of each pinna as **bipinate fronds.**

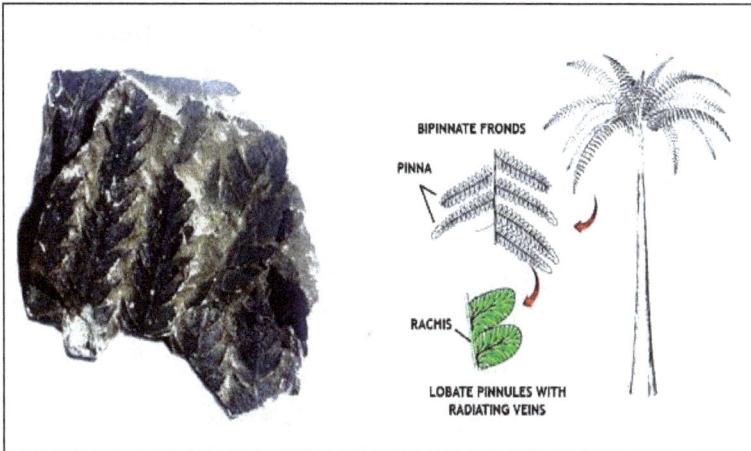

Figure 4.5: *Pecopteris* fossil and reconstruction of the plant and detail of frond

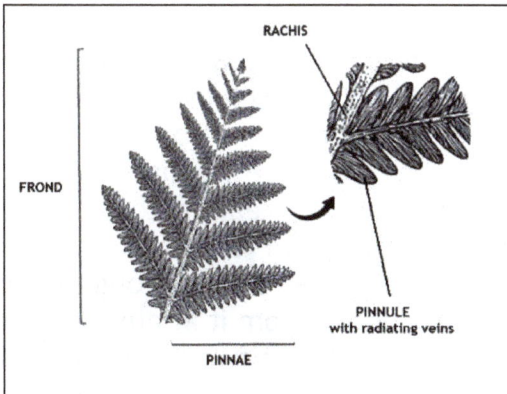

Figure 4.6: Drawing of a fossil fern showing its structure

- *Ginkgo* species is known from the Permian Period to the Pliocene Epoch, but with a relative still living today called *Ginkgo biloba,* or the Maidenhair Tree, which is found in Japan and China. It is a seed-fern, which belongs to Division Ginkgophyta. In the fossil record, it is distinguished by a radiating, hand-like frond or **whorl** of long, spathulate leaves which are sharply cut off at their ends. Veins are long and are parallel to each other.

Figure 4.7: Part of a whorl and a complete whorl of a *Ginkgo* species

- *Taeniopteris* species *such as T. lentriculiformae* from the Latin, *Taenia,* meaning tapeworm and *pteris,* meaning a fern, was common throughout the Mesozoic Era, especially from the Late Triassic Period in the Southern Hemisphere. This fern-like plant was possibly related to the early cycads belonging to Division Pterophyta. It had a broad, ribbon-shaped pinnae and a strong **midrib** with parallel veins coming from it at almost at 90 degrees. Veins often split into two or **bifurcated.**

Figure 4.8: *Taeniopteris* species and a diagram showing bifurcation of veins

- **Linguifolium** species such as *L. tension-woodsii* is known from the Middle to Late Triassic Period, and is a member of Division Pterophyta. It has pinnae which are narrower than *Taeniopteris*, and its narrow midrib is not as pronounced. Leaves are elongate and slightly spathulate with a rounded apex. Veins come off the midrib at about 60 degrees and may be bifurcated.

Figure 4.9: Long leaf of *Linguifolium* and detail showing venation

- **Dicroidium** species such as *D. odontopteroides*. Its name refers to the bifurcation of the rachis di- and the tooth-like pinnules -odonto. It is another seed fern of Division Pterophyta common to the Middle to Late Triassic Period of the Southern Hemisphere. Pinnules are broad along the rachis and are **lobate** in shape and has veins which radiate out from the centre of the base of each pinnule.

Dicroidium elongatum has also been called **Xylopteris**, and is a seed fern of Division Pterophyta common to the Middle to Late Triassic Period of the Southern

Hemisphere. It has long grass-like pinnae with a faint central midrib which branch out from a central rachis.

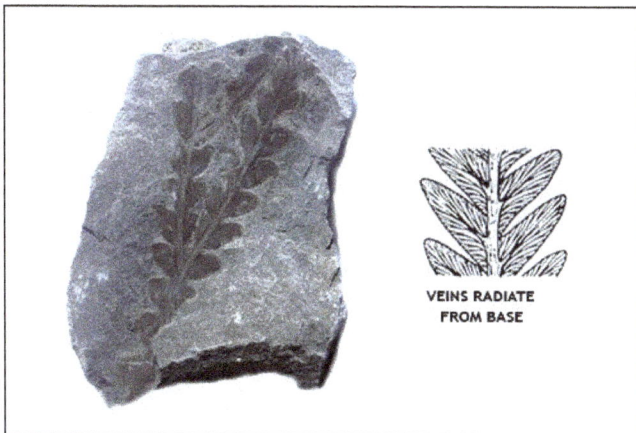

VEINS RADIATE
FROM BASE

Figure 4.10: *Dicroidium odontopteroides* with its typical bifurcated rachis and (at right) detail showing radiating veins

Figure 4.11: *Dicroidium elongatum* (sometimes called *Xylopteris*)

- *Cladophlebis* species such as *C. australis* was a tree fern of Division Pterophyta, common to the Middle to Late Triassic Period of the Southern Hemisphere, but some species of *Cladophlebis* also range into the Early Cretaceous Period. Pinnae have many closely-packed, short oval pinnules alternating on each side of the midrib of the pinnae. These pinnules have broad bases with radiating veins usually bifurcating twice near their ends. A similar plant is *Cladophlebis indica* which has veins which only bifurcate once.

Figure 4.12: *Cladophebis australis* frond (left) and detail (right) showing long pinnae

Online Video 4.1: Southwestern Tasmania resembles the Triassic Period of the southern hemisphere. Go to https://youtu.be/g0t8ouPbd9g

Figure 4.13: Drawing of the double bifurcation in *C. australis*

In the Late Triassic shales of the Southern Hemisphere, is an unknown ancestor of the modern Dicotyledons of Division Angiospermophyta. These are similar to the modern Antarctic Beech found in mudstones from the Middle Eocene Epoch to today. These fossils have the typically oval-shaped leaves with serrated edges and strong network veins. Their seeds have two covers or **cotyledons.**

Figure 4.14: Dicotyledonous fossil from Denmark Hill, Queensland, Australia (left) and a modern leaf from a modern Antarctic Beech, *Nothofagus moorei* (right).

Chapter 5: Invertebrate Fossils

5.1 Introduction

Animals are more difficult to find as individual fossils, but large-scale outcrops such as fossiliferous limestones, provide a richer source because colonial corals and other animals of the original marine reef were preserved together. Animals are less common as fossils than plants because they are:

- larger and more mobile and so rarely become covered with preserving sediment

- often eaten by other animals with few remains

- quickly decayed after death

Animal fossils also leave evidence of their passing such as burrows, footprints and nests. Both plant and animal fossils are often good indicators of their original environment, with animal fossils generally suggesting more detail about their lifestyle and habitat. For example, Foraminifera, minute marine single-celled animals, can be used to reconstruct past climates from the examination of the chemical composition of their shells or **tests**. Fossil animals include both vertebrates, those animals which have a bony spinal column and vertebrates which do not.

5.2 Some Common Invertebrate Fossils

Invertebrates are animals which do not have an internal vertebral column of bony or cartilaginous tissue, so many of them are small and often secrete a protective shell. Many of them are colonial, joining together in large masses formed by the combination of their individual, hard habitats.

Some of the most common Invertebrate fossils include:

- **Foraminifera** named from the Latin for hole bearers, and informally called forams, examples of which include *Marginopora vertebralis*, a tropical, marine, single-celled animal of Kingdom Protozoa. They construct an external test from minerals, usually calcium carbonate from seawater, and hunt by streaming out long tentacle-like structures of cell material called **cytoplasm** as feet-like structures called **pseudopods** through holes in their test. Whilst fossil foram tests have been found as far back as the Cambrian Period (541 to 485.4 million years ago), this species of *Marginopora* has only existed since the Holocene Epoch from about 10.000 years ago, to the present day. Their presence in rock strata is significant, because certain forams can only exist in waters of very certain temperatures. This knowledge can be used as an indicator of the water temperature of that geological time. As marine organisms, their presence in rock strata is also a useful indicator in oil exploration.

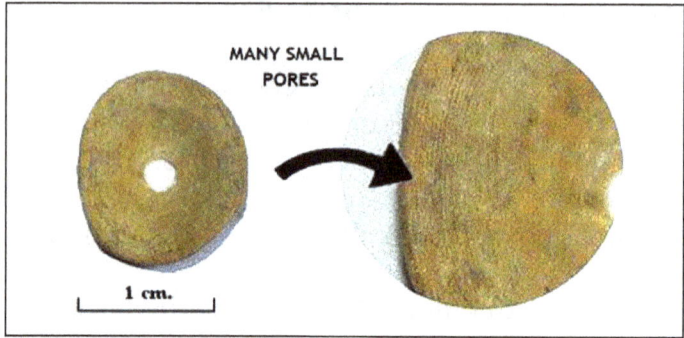

Figure 5.1: Single specimen of *Marginopora* and detail showing the many small pores in its test through which its cell material protrudes as pseudopods (false feet) to gather food.

- **Colonial corals** such as *Favosites* species, are known from the Ordovician to the Permian Periods and formed considerable limestone deposits during the Silurian Period. They are members of Phylum Cnidaria, Class Anthrozoa, Order Tabulata, and consist of colonies of many soft-bodied individuals or **polyps**, living in hexagonal cells called **corallites.** These corallites are made of calcium carbonate, and are joined together in a honeycomb framework with hard outer walls, the **epitheca**, to form the colony or **corallum**. The epitheca are pierced with holes called **mural pores** for circulation of water. The corallites were connected by small pores, and each cell was divided into layers by internal partitions called **tabulae**. They were formed in warm, shallow seas as part of large coral reefs but are now extinct.

Figure 5.2: *Favosites* corallum

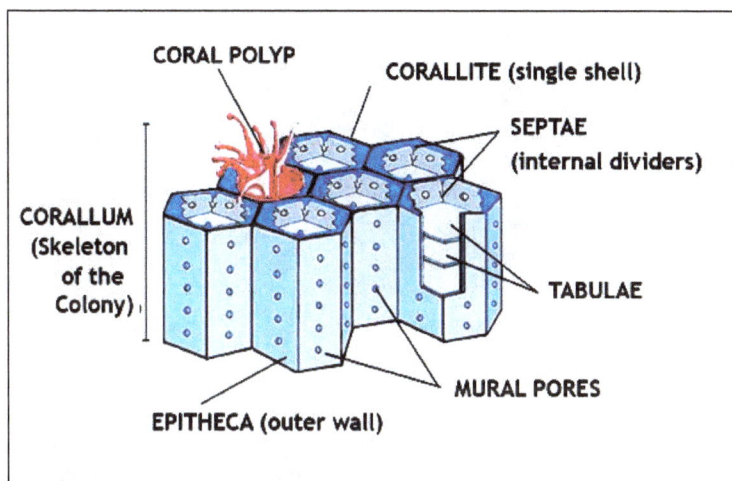

Figure 5.3 Diagram showing the structures within a colonial coral

- **Solitary corals** belong to Order Rugosa and are commonly known as rugose corals. They are usually found as individual corallites, which are larger than those of the **tabulate** corals, in solid masses of limestone. They have a central column or **columella** in the centre of a corallite where the internal partitions or **septae** meet. They also lived in the same warm, shallow, marine environment as the colonial corals and also contributed to reef building.

Figure 5.4: Fossiliferous limestone showing rugose corals. Detail shows a separate corallite with its septae, or internal vertical partitions.

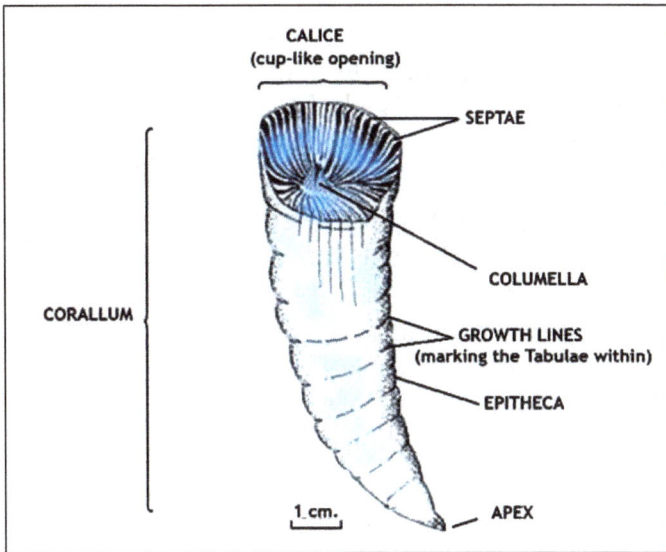

Figure 5.5 Details of the structure of a solitary coral

Figure 5.6: *Ketaphyllum species*, a rugose coral from the Devonian Period

- *Charniodiscus* species was a primitive marine animal named from the locality of its discovery, the Charnwood Forest in Britain. It was probably a lone filter-feeder which anchored itself to the seafloor by its disc-shaped structure at the end of its body which acted as a sucker. It is known from the Proterozoic Period from about 560 million years ago.

Figure 5.7: *Charniodiscus species* possibly in its correct orientation showing the disc-like sucker at its base.

- **Graptolites** are named from the Greek, *graptos*, meaning written, and lithos meaning rock, and are members of Phylum Hemichorda, Class Graptolithina. Known from the Cambrian to the Carboniferous Periods, they were colonial marine animals whose soft-bodied polyps or **zooids**, lived in cup-like structures called **theca**. These cups were arranged along a thread-like branch or **stipe** with the end or **sicula**, possibly housing the initial individual who reproduced to form the others. There was a tube called the **stolon**, within the stripe connecting each zooid and this probably acted as a primitive nervous cord. Some graptolites colonies or **rhabdosomes**, were sessile, that is they were attached to the sea floor, whilst others had a float and drifted on the surface. Graptolites came in many different varieties depending upon the number of their branches.

Figure 5.8: Two species of graptolites in a black mudstone - *Dicranograptus* (left) and *Tetragraptus* (right)

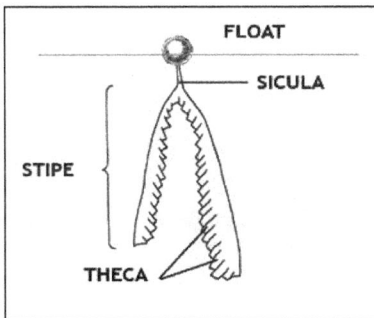

Figure 5.9: A simplified diagram of a floating graptolite

- **Crinoids** are named from the Greek, *krinon,* for a lily, and eidos for form, belong to Phylum Echinodermata, Class Crinoidea. They have been found in rocks dating back to the Ordovician Period, but some species still survive today. are also called sea-lilies because they have flower-like body on a long stem which is attached to the seafloor.

  They live in both shallow and deep water up to 6000 metres depth, and have a long stem made from **calcareous** rings, that is made from calcium carbonate, which supported the top section or **calyx** of feathery arms which are used to filter the water for nutrients. Usually the calyx is rarely found preserved as a fossil, but crinoid stems are often plentiful in limestone deposits.

Figure 5.10: Large Crinoid stems (left) and a hand specimen of crinoidal limestone showing tops of many stems.

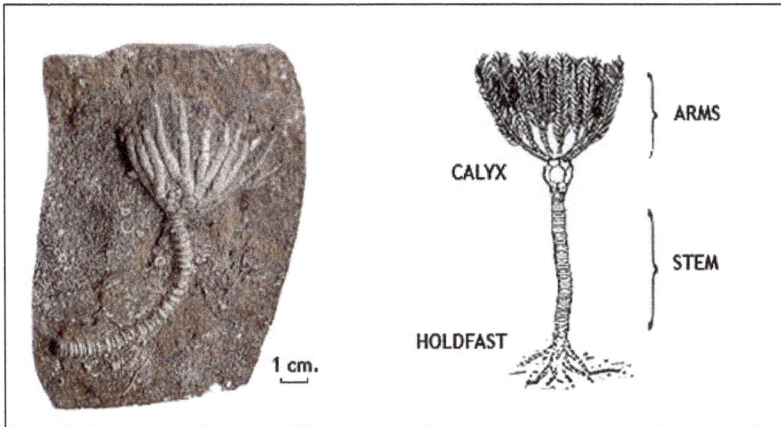

Figure 5.11: *Woodocrinus* species from the Carboniferous Period and a reconstruction of a crinoid showing its main parts

- *Lovenia* species is a marine Echinoderm with a spikey outer skin found in shallow pools and on the sea floor. Fossils of these creatures have been found in marine limestones back to the Miocene Epoch and they still exist today. They are small (2-5cm), bristle-covered, red-pink sea urchins which move about on many small tube feet looking for food. Like most sea urchins, *Lovenia* has a hard outer covering made up of many small, fused calcareous plates in **radial symmetry.** Because their tube feet are only on their underside, this type of Echinoderm is classified as an Irregular Echinoid, unlike the Regular Echinoids which are covered in many tube feet all over.

Figure 5.12: Several dorsal (upper) views of Lovenia woodsia showing plate markings

Lovenia woodsia, the heart urchin, has the classification of:

Phylum: Echinodermata
Class: Echinoidea
Order: Spatangoida
Family: Loveniidae
Genus: *Lovenia*
Species: *woodsia*

- **Fenestella** species is named from the Latin, *fenestra*, meaning little window, for the window-like openings in the mesh of its colonial framework. It is a member of Phylum Bryozoa, and lived from the Ordovician to late Triassic Periods. They are also called Moss Animals or Lace Corals. The colony consisted of small polyps or zooids living within the walls of opposing sides of the small window-like openings called **fenestrules** of the fan-like framework. The sides of the fenestrule are connected by the upper and lower sides called the **dissepiments.**

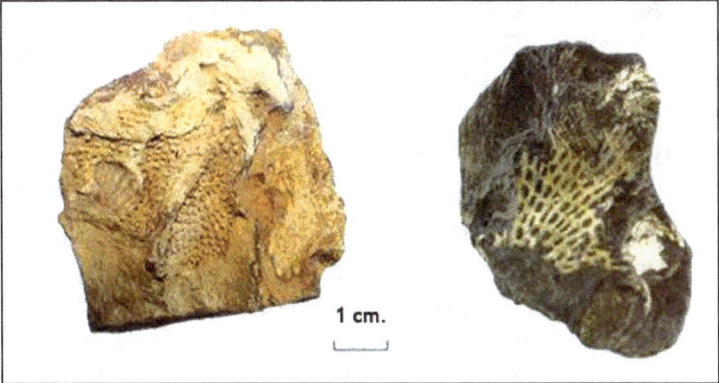

Figure 5.13: *Fenestella species* (left) from the Devonian and an unknown species from the Permian (right)

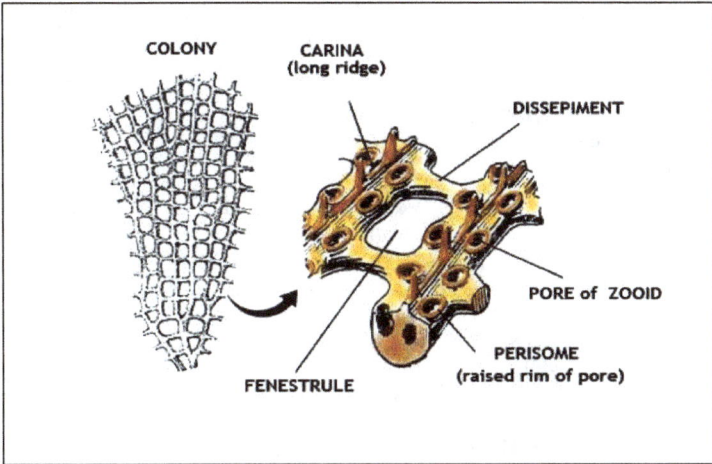

Figure 5.14: Diagram showing the main parts of *Fenestella* species

- **Brachiopods** named from the Ancient Greek words for arm and foot, were marine **bivalves,** having two shells as an upper and a lower shell. They are also known as lamp shells, because they resembled an ancient Greek oil lamp. They belonged to Phylum Brachiopoda, and were known throughout the Palaeozoic Era, especially in the Carboniferous Period; but there is still a species living today called *Lingula species*. Brachiopod **valves** are hinged at the rear end and open at the front for feeding or closed for protection. They have a long, muscular foot called a **pedicle**, which projects from an opening or **foramen**, in one of the valves, the pedicle valve, which is used to anchor the brachiopod to the seafloor. at the edge of the valve is the **lophophore**, a crown of tentacles whose fine hairs, called **cilia**, move to create a water current which enables them to filter food particles out of the water as well being used as for respiration.

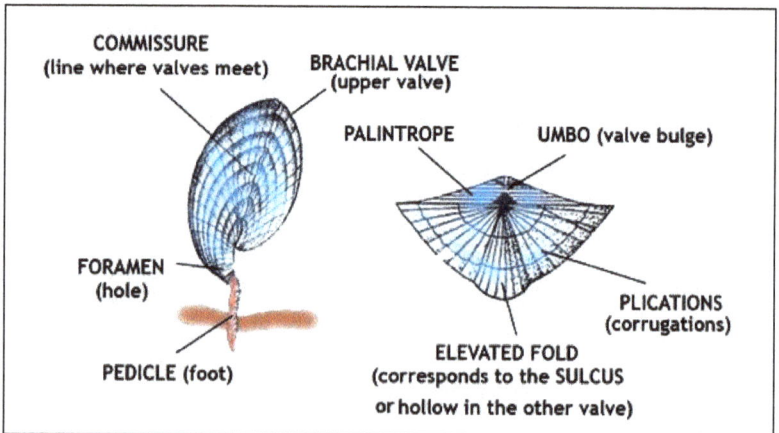

Figure 5.15: Diagram showing the main parts of Brachiopods

Two major groups of brachiopod are recognized, Class Articulata and Class Inarticulata. **Articulate** brachiopods have toothed i.e. articulated hinges and simple muscles for opening and closing their valves. Inarticulate brachiopods have hinges without teeth and a more complex system of valve muscles. One of the most common types of brachiopod belonged to Order Spiriferida – the spirifers. These had elongate, wing-like valves and were common between the middle Ordovician and the middle Permian.

Figure 5.16: Two species of *Spirifer* preserved as a cast (left) and as a mold (right)

- **Molluscs** belong to Phylum Mollusca, and are named from the Latin, mollis, meaning soft. This phylum consists of a very diverse range of animals living in a variety of habitats – marine, freshwater and on land. Some have two shells as bivalves, a single shell as univalves or no shell at all. They include modern marine shellfish, snails, octopuses and squids. They all have a cape of tissue called a **mantle** which is used for respiration and excretion, a long, rasping tongue called a **radula** and a well-defined nervous system. Primitive molluscs probably first appeared in the Cambrian Period. There are seven classes of mollusc:

- Class Aplacophora they lack a shell but are covered with small spike-like spicules, are worm-like and found in deep marine environments e.g. *Falcidens* species.

- Class Polyplacophora are marine with articulated, aragonite (calcium carbonate $CaCO_3$) plates e.g. chitons.

- Class Monoplacophora are deep marine animals with only one simple shell e.g. *Micropilina*.

- Class Scaphopoda are marine and living on the bottom of oceans. They have one long slim shell e.g. tusk shells.

- Class Gastropoda exist in a number of habitats such as marine, freshwater and terrestrial and are univalves or have no shell at all e.g. snails, slugs, limpets, abalone and conch shells

- Class Cephalopoda are marine molluscs which are univalve or lack a shell. They have a prominent head with well-developed eyes and tentacles. e.g. squid, octopuses, ammonites and nautilus.

- Class Bivalva have both marine and freshwater species. All are bivalves and are filter feeders. They were once given the name Pelecyopods and include clams, oysters, mussels, and scallops.

Some examples of mollusc fossils include:

- *Turritella* species named from the Latin, *turritus*, meaning turreted and the diminutive suffix –*ella* for little turret. They are marine univalve gastropods having a coiled, elongated shell and they are known from the Cretaceous Period to the Holocene Epoch.

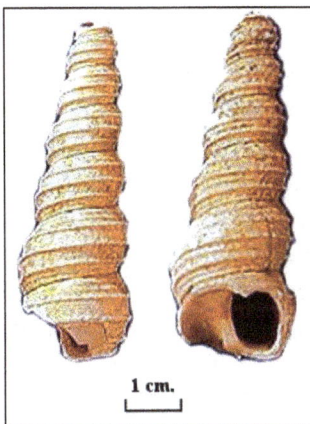

Figure 5.17: *Turritella* species

1 cm.

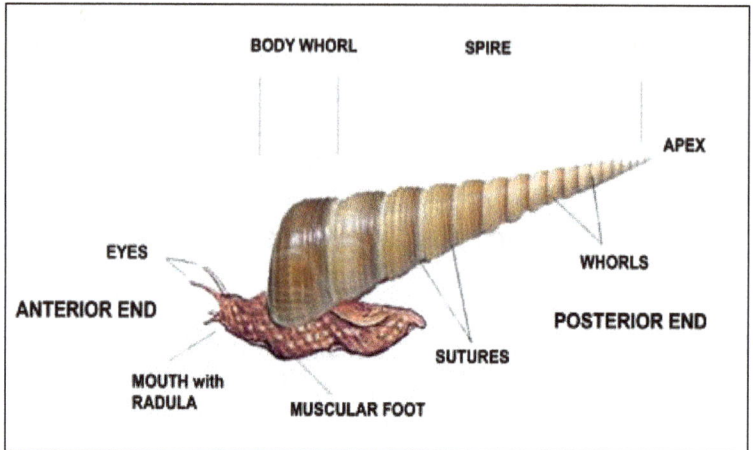

Figure 5.18: Reconstruction of *Turretella* species showing main parts of gastropods

- **Belemnites** named from the Greek, *belemnon*, meaning a dart or arrow, and are known from the Early Jurassic to Late Cretaceous Periods. They were free-swimming marine **cephalopods** closely resembling modern squid. They had a long chambered section called a **phragmocone** with walls made of the mineral aragonite, a form of calcium carbonate, and a heavy surrounding calcite guard called a **rostrum.** At the other end of its body was the head which had a number of tentacles for catching prey.

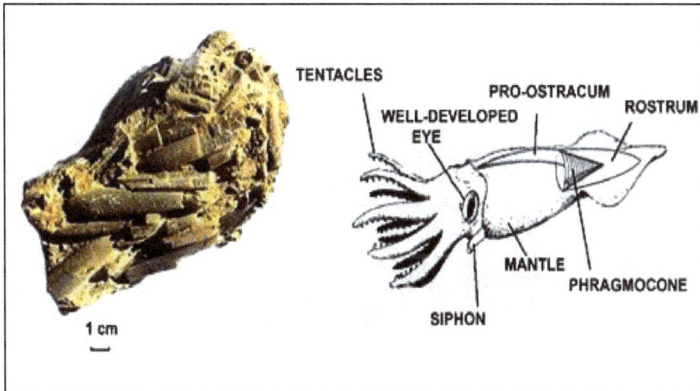

Figure 5.19: A group of fossilized Belemnites (left) and a diagram showing main parts of a reconstructed Belemnite (right).

- **Ammonites** belonged to Subclass Ammonoidea and the name was inspired by the spiral shape of their shells, which resemble tightly coiled ram's horns. These were once called *ammonis cornua* or the horns of Ammon, because the Egyptian god Ammon was typically depicted wearing ram's horns. Known from the Devonian to the Cretaceous Periods, these cephalopods usually had a flat spiral shell or **planispiral** shell, and lived in a marine environment, probably as free-swimming and floating animals like modern-day nautilus. The ammonites had a chambered shell with new chambers (or **camerae)** being added as the animal grew and moved into the final chamber. These chambers were connected by a tube called the **siphuncle** which allowed the removal of water which also assisted in changing the buoyancy of the ammonite. They ranged in size from a few centimetres to over two metres in height.

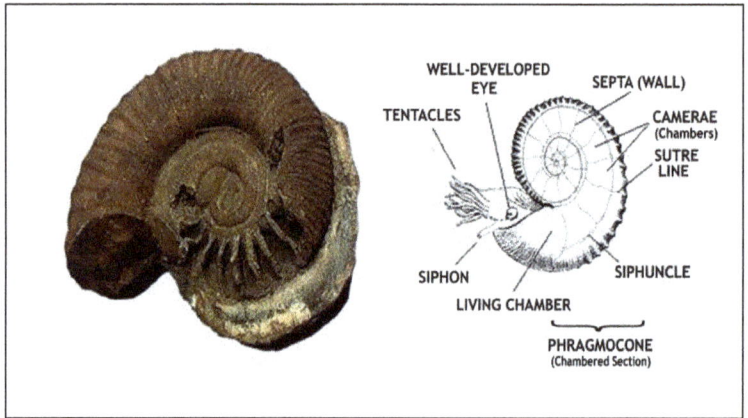

Figure 5.20: Ammonite specimen (left) and diagram showing main parts of the animal (right).

- *Anadara* species was a marine bivalve mollusc which lived between the Cretaceous to the Quaternary Periods. Specimens show the **adductor scars** where the muscles for opening and closing the valves were attached and the **pallial line** where the mantle was attached to the valves (shells).

Figure 5.21: A modern mollusc showing how the valves opened in a horizontal position unlike the brachiopods which opened in a vertical position

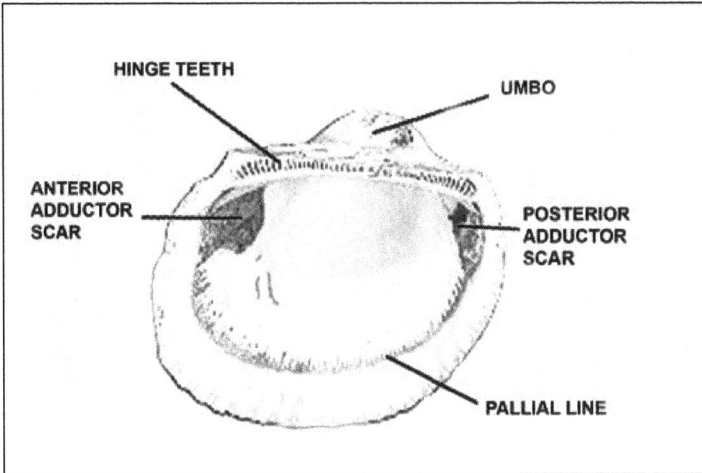

Figure 5.22: Left valve of *Anadara* species and diagram showing its main parts

- **Trilobites** were marine members of Phylum Arthropoda, Class Trilobita, and as their name suggests, they had a hard **exoskeleton** divided into three parts or lobes. Their body was also divided into a **cephalon** or head, a **thorax** or body, and a **pygidium** or tail. The head also had a bulbous forehead or **glabella**, and some early trilobites had **facial sutures** or grooves. They also had many pairs of jointed legs which held gills used for respiration and many species had complex compound eyes. They were known from the Cambrian Period, where they dominated, to the Permian Period. They existed in a great variety of size and form with some scavenging in the muds of the sea floor and others were mobile carnivores. They more resembled modern arachnids such as spiders, than crustaceans such as crabs, crayfish.

Figure 5.23: *Xystridura* species from the middle Cambrian and *Redlichia* species from the lower Cambrian

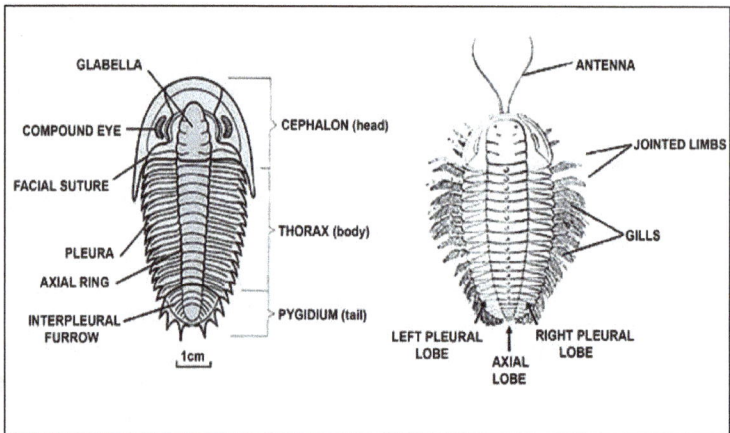

Figure 5.24: Diagrams of *Dalmanites* species (left) and *Triarthrus* species (right) showing their main parts

Chapter 6: Vertebrate Fossils

6.1 Introduction

Vertebrates are those animals which have a vertebral column of linked bones surrounding a notochord of nerve tissue. In some primitive types, such as sharks and rays, the vertebral column is made out of cartilage and not harder bone. Vertebrate fossils are not as easy to find as invertebrate fossils due to their mobility, however the resistance of bones to decay has provided a rich collection in those few localities where there has been a sudden burial followed by limited disturbance. In addition, mobile and scavenging animals sometimes leave behind traces of their activities including burrows, nests, eggs and tracks for the fossil record. Many have **bilateral symmetry**, that is, they have identical features on either side of a line drawn down the long axis of their bodies.

6.2 Some Common Vertebrates

The most common vertebrates include:

- **Fish** which were perhaps the oldest true vertebrate fossil and the jawless fishes first appeared in the Cambrian Period. Species such as *Haikouichthys* were small, tapered, streamlined animals with eyes, gill arches, a **notochord**, and rudimentary vertebrae. These fossils were found in China, although there may be an even older vertebrate in the Precambrian rocks of South Australia.

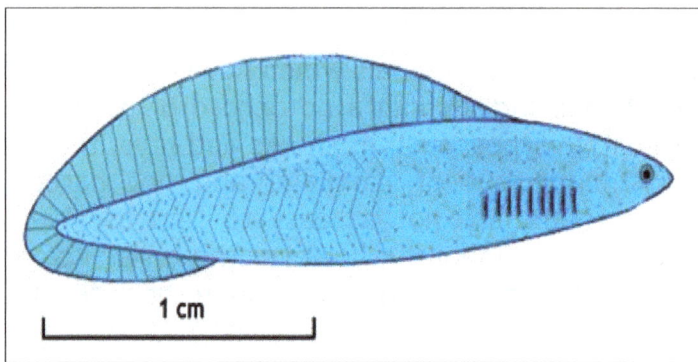

Figure 6.1: A reconstruction of the early fish, *Haikouichthys* species

Some of the earliest known fossil vertebrates were the heavily armoured fish found in rocks of the Ordovician Period. These gradually evolved into the various classes of fish found in the fossil record and in today's oceans and streams. Some of the earliest fishes were the **placoderms**, named from the Greek for plate skinned.

Figure 6.2: Skull of the placoderm *Dunkleosteus sp.* of the Late Devonian Period. This skull is about 1.5 metres long with the fish being about seven metres long.

Figure 6.3: *Knightia species* was a fossilised bony fish from the Eocene Epoch showing many similarities to modern fish.

- **Amphibians** developed in the Devonian Period from the primitive air-breathing fish, similar to today's lungfish. These were able to slowly colonize the land, thus becoming the first terrestrial vertebrates - the amphibians. The young of these animals developed from soft eggs laid in water and in their early larval stage of development they had external gills and had tails to aid them swimming.

Figure 6.4: Skull of *Xenobrachyops species*, an amphibian from the Triassic Period.

Today's amphibians, the frogs, toads and salamanders have these features during their larval stage as well as smooth moist skin and well-developed eyes. As they matured, amphibians developed a simple lung and limbs and moved onto the land but returned to water to reproduce and lay their many soft eggs. It is thought that early amphibians also had a similar life cycle.

- **Reptiles** evolved from those amphibians which developed more efficient forms of reproduction by the ability to lay eggs on land. Animals which do this are referred to as **amniotes,** and they all have eggs which have a dry, leathery shell which was more protective and prevented dehydration. This gave them a better chance of survival. Other features which assisted the early reptiles to a land environment include better locomotion upon strong legs. More primitive reptiles, especially those which were aquatic, like today's alligator and crocodiles, had their limbs coming out from the sides of their bodies but larger reptiles developed legs which were situated below their trunk and this enabled them to stand upright and have better mobility. This in turn was assisted by the development of a metabolism better suited for the changing temperatures and dryness on land. The reptiles which developed from these early land amphibians first appeared in the Carboniferous Period and rapidly diversified in the Permian Period and Mesozoic Era.

Figure 6.5: A reconstruction of *Dimetrodon,* a mammal-like reptile which dominated the Early Permian

The group of animals, generally called **dinosaurs,** diverged from their early ancestors the archosaurs during the Middle to Late Triassic Period, some 20 million years after the Permian-Triassic Extinction Event. This event is thought to have been caused by an asteroid collision with

Earth and it is estimated that it wiped out over 75% of all land vertebrates, including the *Dimetrodon* species, and 95% of marine species.

Figure 6.6: *Pachypleurosaurus edwardsii* was an early Nothosaur or semi-aquatic reptile with webbed feet. This specimen is about 20 cm. long

Dinosaurs varied greatly in form and size, from the small *Xixianykus* of about 50 centimetres long to the huge **sauropods** or long neck, lizard footed dinosaurs such as *Brachiosaurus*, which was over 26 metres long. Some bone specimens found more recently, suggested that there were similar creatures even larger in size. Dinosaurs and related creatures evolved to live in a variety of habitats and had many types of adaptations.

Some of these earlier creatures took to the air such as the **pterosaurs**, meaning winged lizard, which existed from the Late Triassic to Early Cretaceous Periods. They had hollow bones and a thin membranous wing which stretched between their ankles and an extended fourth finger. They were competent fliers and gliders inhabiting cliffs near the sea.

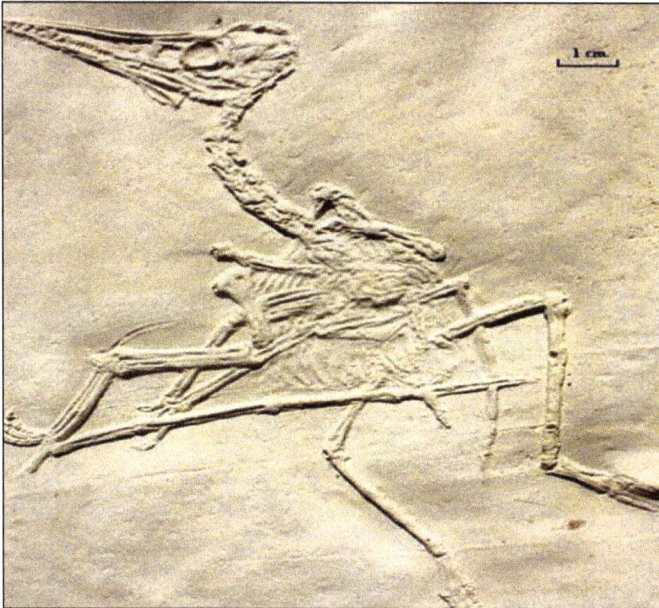

Figure 6.7: Replica of a Jurassic Period Pterosaur from the Solnhofen Limestone, Bavaria, Germany

- Birds first appeared in the Late Cretaceous and probably evolved from **theropods,** a name from the Greek meaning beast feet, which were dinosaurs living during the Jurassic Period. The theropods included some of the largest carnivorous animals which ran on two legs, including Tyrannosaurus rex.

Birds evolved from smaller theropods which also had a wishbone structure called a **furcula**, in the skeleton of their chest which assisted in wing movement. They also had well-developed lungs, air-filled bones, laid hard-shelled eggs in nests and some had feathers. Some birds survived the extinction event that occurred 65 million years ago, and their descendants continue the dinosaur lineage to the present day.

Figure 6.8: Replica of *Archaeopteryx* species, an early bird of the Jurassic Period from the Solnhofen Limestone, Bavaria, Germany

- Mammals named from Latin, *mamma* for breast, first appeared in the Triassic Period and probably developed from the early mammal-like reptiles. This evolution probably continued through the **therapsids**, animals with legs under their bodies and with teeth and skull features similar to modern mammals. These probably first appeared as the **cynodonts** or dog teeth animals, and which then evolved to the modern mammalian orders in the Cainozoic Era after the extinction of the dinosaurs at the end of the Cretaceous Period.

Figure 6.9: A reconstruction of an early Cynodont.

Chapter 7: Evolution and Extinction

7.1 Introduction

By the 20th century the modern theory of **evolution** had been established from a combination of some of the basic principles of Charles Darwin's classical theory of evolution by natural selection, and the modern science of genetics. In simple terms, this theory suggests that those living things which are better able to adapt to their changing environment will pass their characteristics on to the next generation through their genetic characteristics. It is thought that the slight changes in the genetic structure of organisms occurs through random **mutations** or permanent changes in the organism's cellular nucleus, due to external factors such as natural radiation, some chemical influences and possibly some internal influences.

7.2 Evidence for Evolution

Evolution as one of the major cornerstones of modern science has been based on evidence not only from palaeontology, but also from biological sciences, anthropology, psychology, astrophysics, chemistry, physics and other scientific disciplines. The main evidence for this modern theory comes from:

- Common genetic codes in all cells which contain a cell nucleus. In every Kingdom from bacteria to human body, cells contain **DNA** (deoxyribose nucleic acid). This is the basic chemical combination of coded molecules arranged in a double helix or spiral structure which exist in the **genes** or packets on the **chromosomes** within the nucleus of cells. Much of this genetic coding is shared between many organisms. For example, approximately 96% of the human genetic code is shared with chimpanzees and large percentages with other mammals. This suggests that humans and other animals may have had a common ancestor in the past and that the amount of difference between our genetic codes, or genomes, corresponds to how long ago our genetic lines have diverged.

- Where there is a good, continuous fossil record in the rocks, the general trend of a group of organisms can be seen, usually with a gradual transition from one form to another over geological time. The sequence may also show extinctions of some life forms which could not adapt. Some of the best known and observable changes in the fossil record come from animals commonly preserved in large quantities over a long time period. For example, Trilobites show a great number of changes in body structures adapted to different marine environments. The trend of these changes varies from one species and environment to another and may go from simple to complex or in the reverse order depending upon changes in the environment. In trilobites these trends include changes in the number of body (thoracic) segments, size of the eyes and grooves and shape of the head (**cephalon**) and number of body spines.

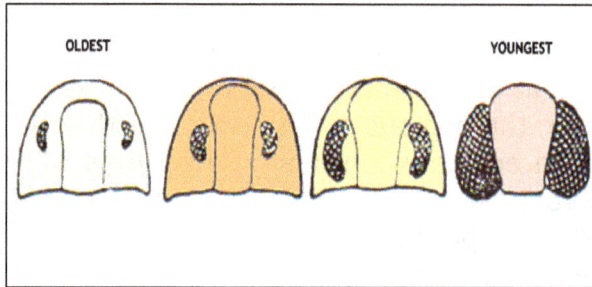

Figure 7.1: A simplification of some of the evolutionary changes in the cephalon or head of one group of Trilobites

- Common features in embryos are characteristic of Phylum Chordata. When the newly-formed embryos of all the species in this group begin to develop, they all have similar features such as tails, and specific anatomical structures involving the spine.

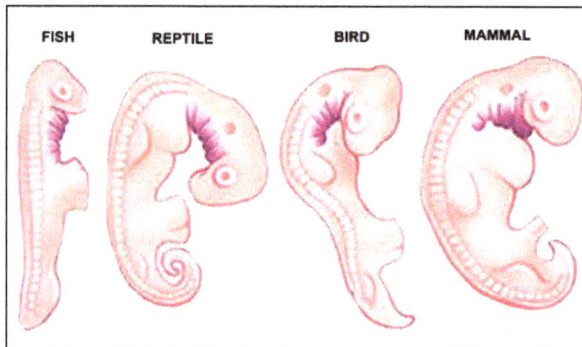

Figure 7.2: Comparative structures of chordate embryos

- Bacterial resistance to new antibiotics is developed by many bacteria species in response to treatment by these antibiotics. The bacteria often show an evolutionary development into new strains which will survive the antibiotic. This is because a few of the original population of bacteria will have those characteristics which will enable them to withstand the new antibiotic. However, some bacteria may develop resistance levels caused by random changes to their DNA as a mutation, which can then be passed onto the next generation whilst the rest of the current population dies out. This process is observable by microbiologists who can not only study the genetic code of bacteria but also observe the changes over many populations as the life span of bacteria populations is very short.

- Comparative anatomy is one of the major threads of palaeontology and physiology. The observation that many organisms, especially in the vertebrates, had many common features of their anatomy was first noted by **Georges Cuvier** (French: 1769-1832). For example, the features of the vertebrate **pentadactyl limb**, that is their appendages such as a leg, arm, flipper or wing, consists of a single large bone, the humerus, connected at one end to the shoulder of the vertebrate, then two smaller bones, the radius and ulna, connected to this bone, and then a hand of five (hence the term *Penta*) fingers or equivalent, of smaller, jointed bones. This structure can be observed in the appendages of humans, whales, bats and other mammals. In addition, modifications of these structures are also seen in amphibians, birds and some fish, especially in some of the more ancient forms.

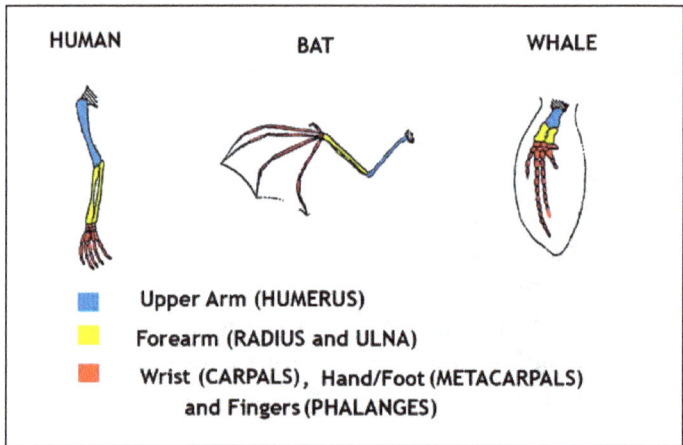

Figure 7.3: Comparative structures of the pentadactyl limb

As an example, consider the evolution of the modern horse (genus *equus*). In the Early Eocene, about 50 million years ago, an animal which had a jaw bone resembling that of a modern horse appeared. This was *Eohippus*, from the Greek for Dawn Horse. Anatomists noted the similarity of this jawbone to that of modern horses, and could estimate the size of *Eohippus* by using similar ratios from the skeleton of the modern horse to construct a scaled-down skeleton for that of Eohippus. After this reconstruction, *Eohippus* was found to be about the size of a medium-sized dog or about 40 centimetres tall.

More bones of *Eohippus* showed a typical pentadactyl limb, but with four toes on its foreleg and three on its hind leg rather than the fused toes as a hoof in modern horses. The rock types, other fossils found with the *Eohippus* bones, and the sedimentary features found in the rock itself, suggested that this first horse probably lived in a wet, forest environment. Over time, other horse-like fossils demonstrated that *Eohippus* evolved through the Cainozoic Era to become adapted to more open, dry plains, slowly becoming the modern genus, *Equus* via a number of intermediate forms:

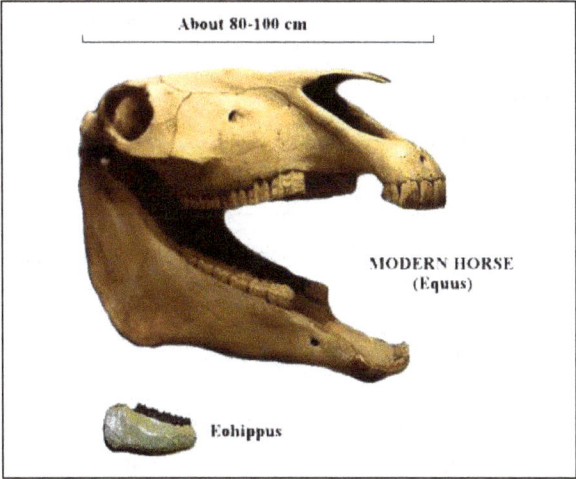

Figure 7.4: Jawbone of *Eohippus* (bottom left) and the skull of a modern horse

EPOCH	GENUS	SKULL	FORE LIMBS	HEIGHT	ENVIRONMENT
Pliocene	*Equus* (Latin for Horse)		One toe	130-180 cm	Grass plains
	Pliohippus (Greek: Pliocene Horse)		Side toes not touching ground	125 cm	
Mid-Miocene	Merychippus (Ruminent Horse)		Three toes	89 cm	Open Grasslands
Oligocene	Mesohippus (Middle Horse)		Side toes touching ground	60 cm	
Mid-Late Eocene	Orohippus (Mountain Horse)		Four toes	About 40 cm	Grass & tougher vegetation
Early Eocene	Eohippus (Dawn Horse)		Three toes on hind limbs		Wet Forests - ate leaves & fruit

Table 7.1: Showing the possible evolution of Equus

7.3 Extinction and Survival

Species and even larger groups evolved, some remaining as **extant** organisms still surviving today, and some have become extinct. In examining the fossil record, palaeontologists have noted that at certain times, some species have simply disappeared. This could be due to lack of fossilization and sedimentary processes, but these extinctions have been more extensive and widespread than any localized effect could have on the reduction of fossils. Most scientists now agree that these sudden disappearances where due to catastrophic events which were sudden, and involved the entire Earth. They are referred to as mass extinctions.

Scientists have identified five time periods which may have been mass extinctions:

- Ordovician-Silurian extinction event which occurred at the end of the Ordovician Period. It may have consisted of two separate events which destroyed a large number of marine species, especially trilobites, graptolites and brachiopods. During this time there was a sudden drop in sea-level as glaciers formed, followed by a time when there was a sudden increase in sea- level as glaciers melted.

- Devonian-Carboniferous extinction event which occurred at the later part of the Devonian Period, when a series of extinctions eliminated a significant number of species, especially those of shallow seas such as many of the coral species. It is possible that global cooling and some sea-level reduction may have triggered these extinctions.

- Permian-Triassic extinction event which occurred at the end of the Permian Period was possibly the Earth's largest extinction, killing a majority of all species. This included both terrestrial and marine species, including many insect species and the mammal-like reptiles. It did, however, allow for the rise of the archosaurs – the ancestors of the dinosaurs, birds and crocodilians in the Triassic Period.

- Triassic-Jurassic extinction event which occurred at the end of the Triassic Period and eliminated most of the archosaurs, therapsids, and large amphibians. This enabled the dinosaurs to become the dominant animal group on land. About this time there was a great production of flood basalts - eruptions of great volume with little ash which spread over large areas and contributed to an increase in global temperature. It is possible that this may have triggered this mass extinction.

- Cretaceous-Lower Tertiary extinction event at the end of the Cretaceous Period. This is the well-known K-T Extinction, from the German, *Kreide*, meaning chalk, and Tertiary. It is the extinction which removed all of the non-avian dinosaurs and assisted in the rise of the mammals and birds as the dominant vertebrate animals. There have been several theories put forward for the cause of this event; most notably that of an asteroid collision with supporting evidence coming from the amount of the chemical element **iridium**. This is a natural chemical element which is found only in the Earth's mantle and extra-terrestrial objects such as meteors and asteroids. It is also found world-wide as a fine layer within rocks of the time of this extinction. Moreover, several large impact craters have been found, one 180 km in diameter dating to the Late Cretaceous buried beneath sediments of the Yucatán Peninsula near Chicxulub, Mexico and a second, smaller and slightly younger crater, at Boltysh in Ukraine. **Tektites**, small broken fragments typical of such impacts, have also been found in deposits associated with the time of this extinction. As well as the indications of impact, including those caused by the huge fireball and tsunami which would have occurred at that time, such a collision would have also caused a major reduction in the amount of sunlight needed for plants and therefore the rest of the food-chain.

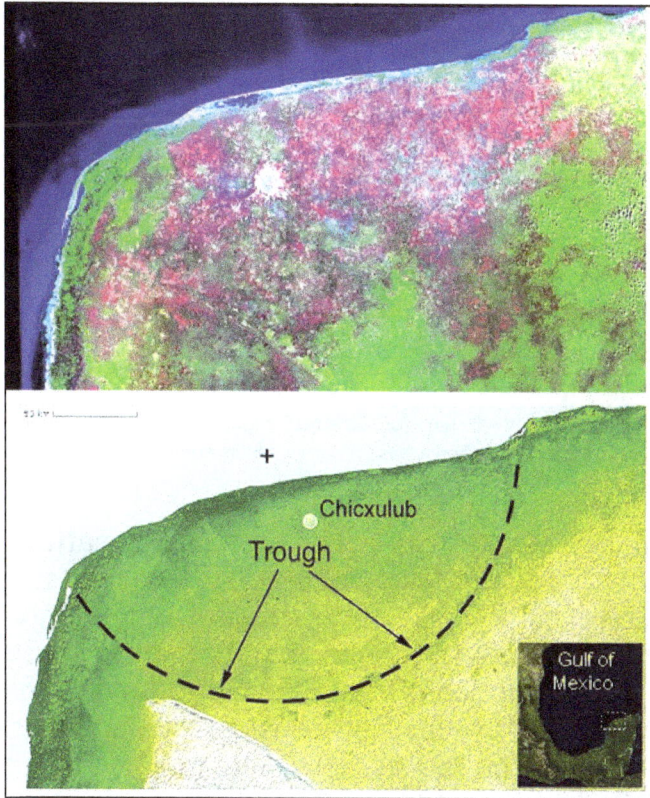

Figure 7.5: Maps showing the possible impact zone of the Chicxulub bolloid - Space Shuttle Radar photo (above) and general location map (below).

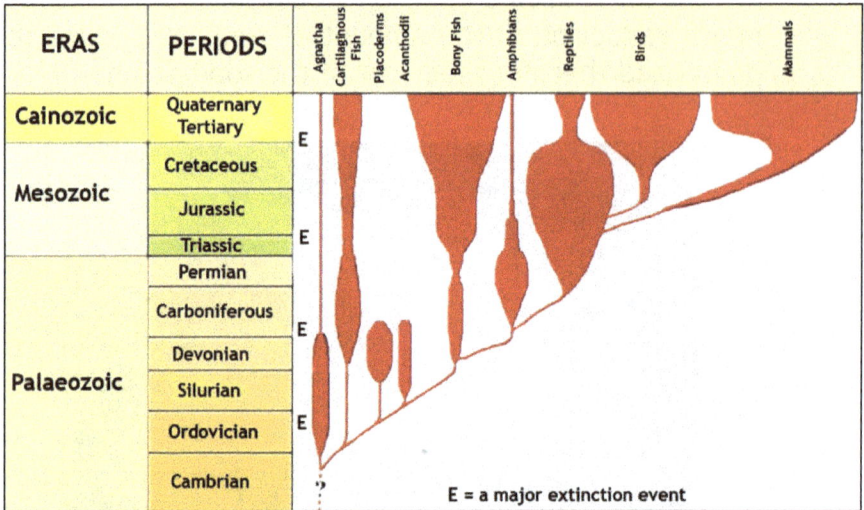

Table 7.2: A simplified chart for some major animal groups showing their development and some extinctions. Red areas are an approximation of the size of each major group.

Whilst this impact theory gives a good explanation of why such extinction could occur, some palaeontologists have doubts that the time of impact at these sites actually coincides with the time of the extinction event. Another theory suggests that large-scale eruptions of basalts at that time in the Deccan of India, would have poured considerable amounts of carbon dioxide into the atmosphere and produced a **greenhouse effect**, which would have warmed the entire planet.

A third theory suggests that sudden movement of the Earth's crustal plates could have cause major disruption to the habitats of the dinosaurs as well as a major climate change in some localities. It is conceivable that there may have been several factors at this time with the asteroid impact possibly being the **tipping point** which caused the Earth's climate to change to such an extreme that mass extinctions occurred.

7.4 Extinctions and Climate Change

The Earth's climate has gone through many changes in climate. The fossil record shows a variety of evidence for this, including periods of glaciation, desertification, high humidity, high temperatures and variations in the percentages of oxygen and carbon dioxide in the air. Throughout the Palaeozoic and later Eras, organisms have either adapted to meet these conditions or have died out.

There have been several distinct periods of cold climate, with large ice sheets growing across Europe and North America, also known as the Ice Ages. Evidence has come from the widespread glacial debris scattered across the Earth's surface during these different geological events and isotopic ratio analysis of oxygen in fossils. Oxygen has two main **isotopes**, ^{16}O and ^{18}O – with a higher concentration of Oxygen ^{16}O in ice. Drill core analysis of marine sediments and of ancient ice sheets in Greenland and Antarctica, have also provided evidence for more recent ice ages.

There have been at least five major ice ages or glacial periods, identified in the Earth's past with relatively warm periods or Interglacial Periods in between in which the higher latitudes of the northern hemisphere were ice free. The main Ice Ages which have been recognized are the:

- Huronian Ice Age which occurred in The Proterozoic Eon about 2100 million years ago. Evidence for this episode of glaciation comes from very old sedimentary deposits typical of glaciation, such as varve shales and basement scouring, around Lake Huron in North America and also in Western Australia.

- Cryogenian Ice Age which occurred near the end of the Proterozoic Eon about 850 to 630 million years ago in the Cryogenic Period sub-division of this Eon. It was an event during which severe glaciation as ice sheets may have reached very low latitudes.

- Andean-Saharan Ice Age which occurred during the Late Ordovician and Silurian Periods, between 850 and 630 million years ago. Evidence for this event comes from rocks of the Andes Mountains, the central Sahara and Western Africa which were then part of the continent of **Gondwana.**

- Karoo Ice Age occurred in the Carboniferous and Permian Periods between 360 and 260 million years ago and which may have been triggered by plate movements associated with the formation of the supercontinent of Pangea. There were probably two separate areas of glaciation during this time with extension of large glaciers – one in Africa and South America which then joined together, and a later episode in India and Australia which also combined.

- Quaternary Ice Age which started about 2.5 million years ago and is the most well-known of all of the Ice Ages. This produced a large ice sheet which spread across northern Europe down to Spain and another moving across North

America, the Laurentide Ice Sheet, which spread over Canada and northern United States. These events carved out the Great Lakes and the many smaller lakes found in North America, and the many small lakes in the Baltic region of Europe. It is this event which is often referred to as The Ice Age, and the one which probably caused the extinction of many large mammals in the affected regions. The corresponding drop in sea-level also saw the migration of many animals, including early humans, across land-bridges, especially between Asia and America and Asia and Australia. Very detailed ice core studies have shown that this major Ice Age has actually consisted of up to eight smaller glacial periods separated by interglacial periods. This period also includes our current interglacial period which began about ten thousand years ago.

There are several natural controls which may have significant effects on the Earth's climate and which may produce cycles of glacial and interglacial periods. These include:

- Strength of the sun's energy which varies over time with an eleven-year sunspot cycle but scientists are not sure as to the exact causes of these cycles nor the effect that these have on climate.

- Reflection of sunlight from off the earth or **albedo,** from the Latin for whiteness, is the reflection of heat from the Earth, especially from the ice caps. This would normally cool the planet since as the heat is reflected and any loss of significant parts of the ice caps, such as being observed today, would add help to retain heat to add to any other global warming mechanism.

- Changes in the earth's orbit which occurs as a natural process in several long-term cycles initially suggested by **Jens Esmark** (Danish: 1763-1839) in 1824, but confirmed in more detail mathematically in 1920 by **Milutin Milankovitch** (Serbian: 1879-1958). Milankovitch was able to show that the three variations in the Earth's orbit, namely **precession**, the way that the Earth's axis wobbles around the axis through its poles, **obliquity** or the tilt of the earth's axis, currently 23.44 degrees and decreasing, and **eccentricity** or changes to the elliptical shape of the Earth's orbit, were closely similar to the cycles of the major Ice Ages. This correlation suggested strongly that Earth orbital variations were a major contributing factor in causing or ending ice age events.

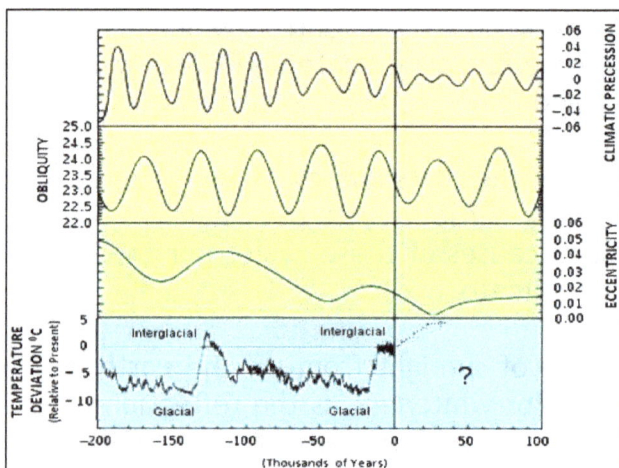

Figure 7.6: Graph showing Milankovitch Cycles for the Earth's orbit and Ice Age estimations. (Modified after Berger & Loutre, 1991, Petit et al., 1999 and others)

- Volcanic eruptions on a massive scale can release large amounts of dust, carbon dioxide and sulfur dioxide into the atmosphere. Dust can be carried long distances around the Earth within high altitude currents and obscure sunlight, lowering the temperature of the Earth to the point where some plants and animal species cannot survive.

Figure 7.7: A huge eruption cloud from Sarychev Volcano, (Kuril Islands) as seen from the International Space Station. Note the pyroclastic flow at the bottom right. (Photo: NASA)

- Meteorite or asteroid impacts such as that described earlier in reference to the mass extinction at the end of the Mesozoic Era would also blanket the Earth with dust causing a reduction in temperatures.

- Extreme greenhouse effects due to the increase in water vapour, carbon dioxide and methane gases within the atmosphere. These gases absorb heat from the Sun and increase the overall temperature of the planet. The temperature of the atmosphere builds up because much this absorbed heat has wavelengths which do not escape through the atmosphere back into space – very much like trapped heat in a glasshouse. Natural emitters of these gases include: living things exhaling water vapour and carbon dioxide during respiration; methane from animals as waste gases; burning of carbon-based fuels to give steam and carbon dioxide; volcanoes erupting water, carbon dioxide and some methane; and decay of organic material such as vegetation which produces methane.

A large amount of methane is derived naturally from the decaying processes. It is also trapped in within vast areas of **permafrost** which is the frozen soil of high latitude countries such as in northern Europe, Russia and Canada. Methane is also emitted from human-based activities such as coal mining, the extraction and use of natural gas and biomass fuels, livestock production and some forms of agriculture, especially rice production.

With increased global temperature, there is a real threat from the vast amount of methane gas trapped in the permafrost across the arctic continents. Methane is approximately 25 times more potent per molecule than carbon dioxide as an absorber of heat over a 100-year period. Methane trapped in the Arctic tundra comes primarily from microbial decomposition of organic matter in soil that thaws seasonally. This methane naturally seeps out of the soil over the course of the year, but global warming could release of even larger emissions from the permafrost. One interesting fact discovered by NASA scientists recently was that more methane seems to be emitted during the colder months in the Arctic when the surface ice acts like a miniature glasshouse and allows the microorganisms below to emit more methane from the trapped organic matter there. Previously this winter emission was not considered in climate change modelling.

Water vapour, whilst a significant greenhouse, does not stay in the atmosphere very long and methane gas, whilst a better absorber of heat, is currently less abundant in the atmosphere than carbon dioxide, which is considered to have the major heating effect.

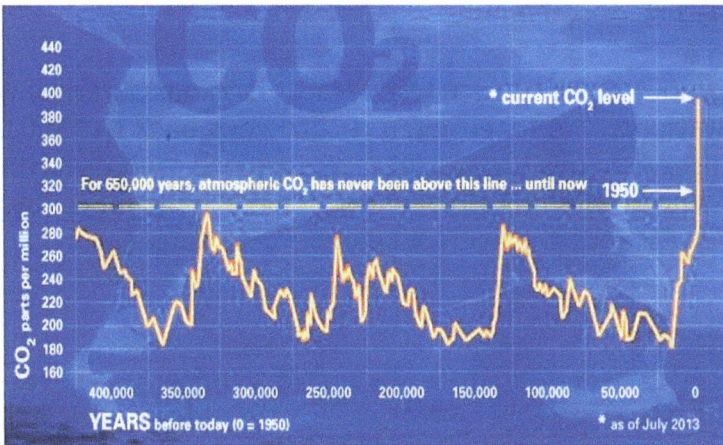

Figure 7.8: Graph showing increased carbon dioxide levels in the atmosphere

There is considerable scientific evidence from many international government organizations, such as the United Nations, NASA, NOOA, CSIRO, and the Royal Academy and so on, which suggest that the Earth is warming and that the natural processes which do this have been supplemented on a larger scale by human activity. Evidence for this induced global warming due to the activities of humankind or **anthropogenic** causes, comes from:

- Ice drill cores from Greenland, Antarctica and elsewhere which give a great variety of data about carbon dioxide levels, oxygen levels, atmospheric dust and temperature as well as other climate information.

- Rock drill cores in sedimentary rocks giving palaeo-environmental data and ages as well as the fossil record and nature of sediment.

- Land and sea surface temperatures as measured by weather stations. The longest sea record goes back to 1850 and the last decade is the warmest since these records have begun. Some problems with heat pollution on land may affect these results, but these errors can be held constant.

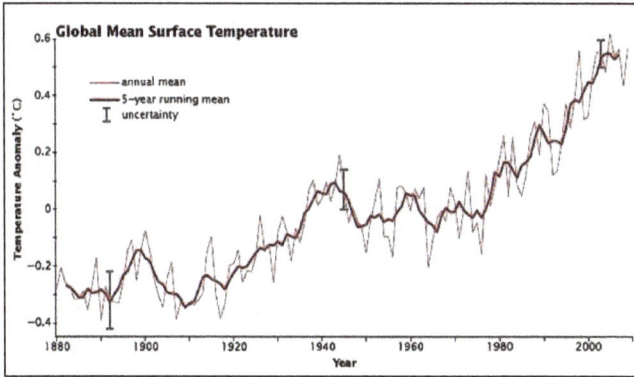

Figure 7.9: Graph showing increasing global heat content of the Earth's surface (Photo: NOOA)

- Ocean heat content, for which records go back over half a century. More than 90% of the extra heat content from global warming is going into the oceans contributing to a rise in sea level through sea-ice melting and water expansion.

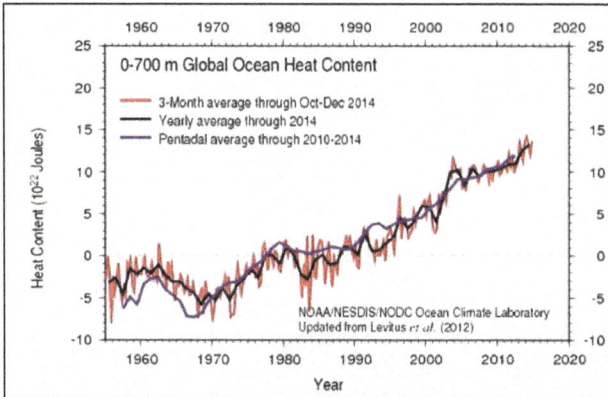

Figure 7.10: Graph showing increasing global heat content of the oceans (Photo: NOOA)

- Measured sea level changes from tidal gauge records going back to 1870, showing that sea level has risen at an accelerating rate.

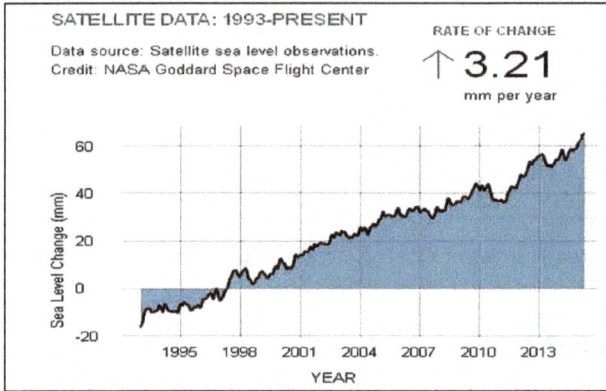

Figure 7.11: Graph showing recent sea-level changes (Photo: NASA photo)

- Glacial retreat for many of the world's glaciers with an overall net loss of ice from glaciers worldwide.

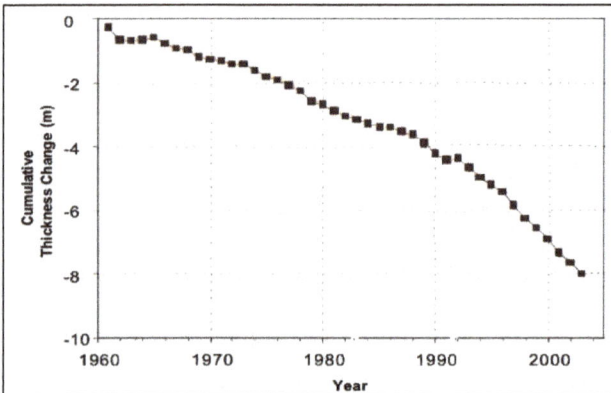

Figure 7.12: Graph showing global cumulative mass change (Modified from that of Mauri Pelto with data from the World Glacier Monitoring Service)

- Northern Hemisphere snow cover has also decreased in recent decades with spring snows disappearing earlier in the year with the total amount of area of snow generally decreasing.

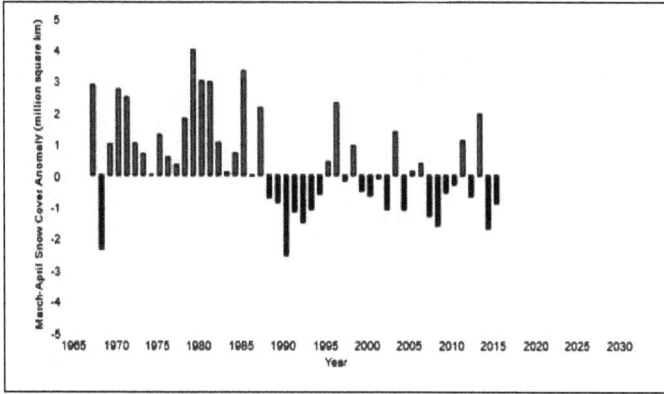

Figure 7.13: Graphing showing average area of northern hemisphere spring snow cover (Photo: NOOA)

- Arctic and Antarctic ice cap reduction is the most dramatic change of all. Satellite measurements from 1979 and shipping records from 1953 show a dramatic loss of sea ice from the high latitude oceans and ice from Greenland and Antarctica.

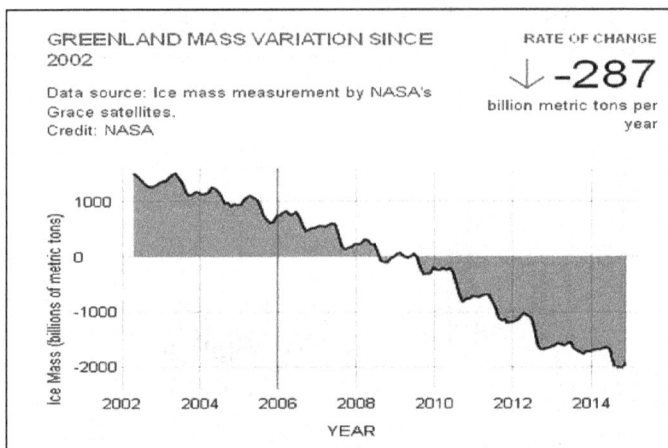

GREENLAND MASS VARIATION SINCE 2002

Data source: Ice mass measurement by NASA's Grace satellites.
Credit: NASA

RATE OF CHANGE

↓ -287

billion metric tons per year

Figure 7.14: Graph showing ice loss from Greenland. A similar but slightly lower rate is occurring in Antarctica (Photo: NASA).

In conclusion: it has been proposed by some scientists that this modern era, the Holocene, is undergoing a new, sixth extinction, the Holocene Extinction, since about 10,000 BC, mainly due to human activity. It has also been suggested that the latter part of our current epoch be renamed the **Anthropocene Epoch,** because of the destructive influence of humankind. The large number of extinctions spans numerous animal and plant species with over 800 extinctions being documented by the International Union for Conservation of Nature and Natural Resources between 1500 and 2009. This extinction possibly includes the disappearance of large mammals starting at the end of the last Ice Age. Such disappearances might be the result of the rise of modern humans and continues into the 21st century and the consequences are yet to be fully understood.

Some optimists have suggested that the Earth as a total environmental system may have the capacity to overcome the effects of anthropogenic climate change and heal itself of all of the negative changes which have occurred due to humankind's activities. This questionable idea has been called the **Gaia hypothesis** named after the ancient Greek Earth goddess. Whilst this hypothesis has not been given much scientific credibility, there have been some occasions where natural processes seemed to have had some minor effect on the negative effects of humankind. For example, it has been noted in the summer of 2012, that some icebergs in Antarctica which have melted due to increased local temperatures have deposited iron oxides which have caused large blooms of algae on the ocean's surface. The iron oxides have come from wind-blown dust from distant land masses. Dropped into the sea, this iron oxide has acted as a fertilizer to increase the population of marine diatoms. There also has been some limited success in artificially "seeding" the ocean using iron sulfate fertilizer. When this is done, there is a sudden increase in the local surface algae population which then quickly dies and sinks to the floor of the ocean. Whilst this has been promoted as a method of sequesting, or burying, atmospheric carbon dioxide, there are still some doubts as to the long-term effects of large amounts of dead ocean-floor algae which is able to returned to the surface after a few hundred years by upwelling ocean-floor currents.

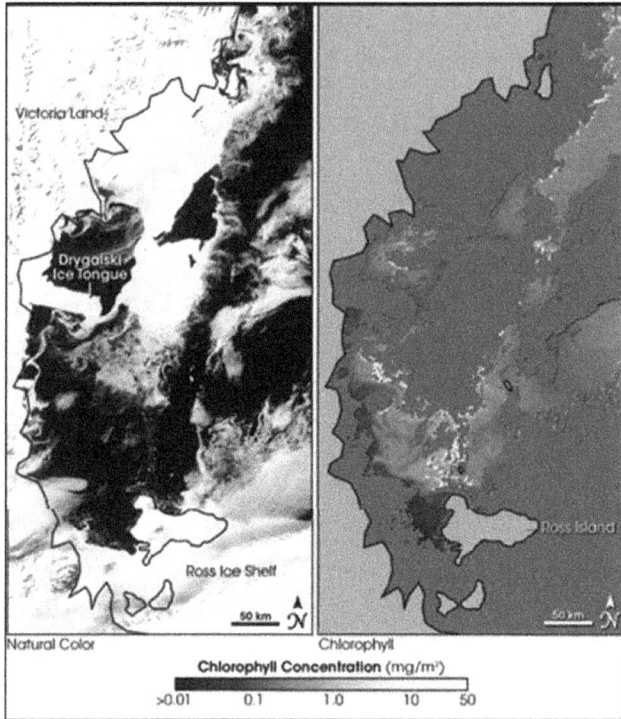

Figure 8.84 Algal blooms observed in Antarctica from NASA's
Aqua satellite (Photo: NASA)

Summary

1. Fossils are the remains of living things which once existed on Earth and which have become extinct or have evolved into modern life.

2. Ancient life can be found preserved by permineralization, replacement, recrystallization, carbonization, impressions, and encasement or by relic traces.

3. Pseudo-fossils are those structures or shapes which look like the remains of living things but are really due to other natural processes such as weathering and sedimentation e.g. dendrites.

4. Fossils, like modern Living Things are classified, or put into groups for study, using the binomial (*two names*) system first developed by Linnaeus in the 18th Century. The main groups are Kingdom, Phylum (or Division for Plants), Class, Order, Family, Genus and Species with individual organisms being identified by their Genus and Species. These are usually written with the species in lower case and both in *Italics* (if printed) and underlined (if hand-written) e.g. *Homo sapiens*.

5. The relative age of a fossil can be determined by referral to a type Locality of a sedimentary rock sequence or generally to other fossils in nearby rocks by noting the order of the fossils within the sedimentary sequence using the Law of Superposition (oldest on the bottom, youngest on the top).

6. More recently, in the latter half of the 20th Century, radiometric dating has used the half-life decay of radioactive isotopes (e.g. Carbon 14) to give an absolute age in millions of years before present for some fossils.

7. The geological time scale was initially constructed in the 19th Century using relative time, later expanded with more sub-divisions and then given approximate absolute ages in the 20th Century.

8. Drawing from data of other sciences and noting the changes in some fossil species in different ages of rock, the modern theory of evolution has been developed. This suggests that slight genetic changes in organisms which make them more suited for survival within changing environments will be passed on to future generations which inherit these changes.

9. There have been five great extinction episodes during which large numbers of plants, animals and other organisms became extinct whilst others have survived. The causes of extinction events have been attributed to such events as massive volcanic eruptions, bolide collisions (asteroids and meteorites), and changes in the Earth's orbit, plate movements and climate change.

10. More recent changes to the Earth's climate have been noted by scientists and the general consensus of opinion appears to suggest that, in addition to natural processes, the actions of humankind have caused a number of climatic changes and extinctions of a large number of species.

Practical Tips

1. Fossils are not easy to find so do your research. Look for locations in quarries or cuttings but take care. Permission must be granted for entering private land and exploration and collection may not be allowed within national parks, reserves nor tribal areas.

2. Fossils are specific to certain past environments so look for plant fossils in finely-bedded shales (coal nearby is often a guide) and marine species in limestone or nearby dark marine shales (oil territory).

3. Very astute observation is needed to find fossils. Sometimes the surface rock weathers differently to the fossil and this may make them more obvious – corals, for example, often appear as slightly raised and a lighter colour on the top of weathered limestone.

4. Some additional preparation is needed in addition to the usual geological pick, field clothing and water. Have a collecting pack (old hiking bag or similar), collecting bags (canvas or cotton is best), newspaper for wrapping delicate specimens and an old wood chisel with a broad blade is useful for splitting shale. A broad felt pen is also useful for labelling the wrapped specimen (see video).

5. Once a good site has been found, look for very fine, darker lines within the laminae of the shales, often these are planes of weakness representing an ancient mud surface and it may contain fossils. Look for sudden colour changes within the rock. Use the sharp blade of the chisel to split the layer with a short, sharp tap with the geological hammer.

6. To preserve delicate fossils (especially in a fine shale), lightly spray the specimen with a fine aerosol lacquer (e.g. hair spray) and allow to completely dry before wrapping. Treat the specimen carefully when transporting it.

7. Make sure that the fossil site is left tidy, any holes dug should be filled in and any gates closed on the way out.

8. Back home, identify the specimen from reputable taxonomy texts or Internet sites then carefully label the specimen using an index card for details and a number (paint a dot about 4 mm. on a hard, flat part of the underside of the rock containing the specimen and when dry use a fine permanent pen to add a simple identification code). On the index card give the name (Genus and species and variety if possible), geological age, location and any other relevant details of each specimen. Place the specimen in a display case or drawer as they will dry out in time and also become dusty in the open.

Multichoice Questions

1. The process of fossil preservation in which a chemical change occurs within the mineral of the fossil converting it to a more stable form is called:

    A. Replacement
    B. Permineralization
    C. Recrystallization
    D. Impressions

2. From your knowledge of ancient life and stratigraphic sequence, the correct order of time Periods from <u>oldest</u> to <u>youngest</u> given below is:

    A. Precambrian – Silurian – Ordovician – Triassic
    B. Silurian – Devonian –Cambrian – Jurassic
    C. Precambrian – Cambrian – Triassic – Carboniferous
    D. Silurian – Devonian – Carboniferous – Permian

3. A fossil species having a wide range of geographic distribution but a small age range is called:

    A. A trace fossil
    B. An assemblage fossil
    C. A limited species
    D. An index fossil

4. From your knowledge of ancient life and evolution, the correct order of fossil types from <u>oldest</u> to <u>youngest</u> given below is:

A. Algae - trilobites - mammals- dinosaurs
B. Algae - trilobites - dinosaurs - mammals
C. Trilobites - dinosaurs - mammals - algae
D. Trilobites - algae - dinosaurs - mammals

5. A fossil plant common to the Late Triassic was *Dicroidium odontopteroides*. The part of this name underlined represents the:

    A. Phylum
    B. Genus
    C. Species
    D. Variety

6. The Law of Superposition is important to the study of fossils because it:

    A. Provides an absolute age range
    B. Provides a relative age range
    C. Provides an index age range
    D. Provides a geological age range

7. The following graph shows fossil extinction against time:

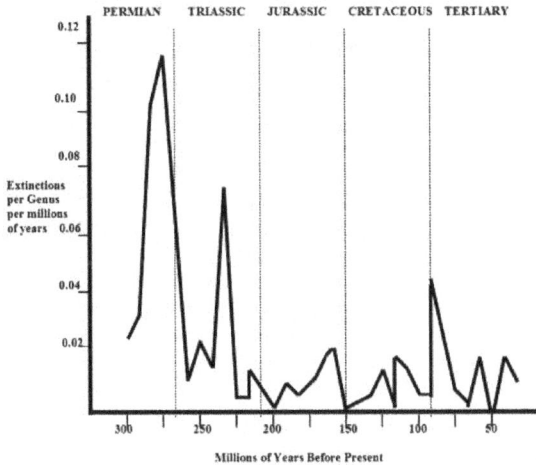

PERMIAN   TRIASSIC   JURASSIC   CRETACEOUS   TERTIARY

Extinctions per Genus per millions of years

Millions of Years Before Present

From this graph, it is probable that:

    A. There is no clear connection between time
        and
        extinctions in the fossil record
    B. There was a major extinction at the end of
        the Jurassic Period
    C. There appears to be a significant extinction
        every 26 million Years
    D.   The average number of extinctions seem to be
about
        0.03 extinctions per genus per million years.

8. The following descriptions of some fossil plants were given
in a research paper:

SPECIMEN A - strong rachis, bifurcated, with alternating lobate pinnules of about 5mm across at base. Base is broad but narrows out to a rounded apex. Veins radiate from a point in the centre of the base.

SPECIMEN B - single, long lanceolate frond about 10 cm long and 1 cm across. Sides are parallel until near the apex which comes to a point. The midrib is very strong and veins are almost perpendicular to it and bifurcate at their ends.

From your knowledge of palaeobotany, these specimens would most likely be identified (respectively) as:

A. *Dicroidium* and *Taeniopteris;*
B. *Pecopteris* and *Glossopteris;*
C. *Ginkgo* and *Lepidodendron;*
D. *Xylopteris* and *Cladophlebis.*

9. The following chart shows the age range for some common fossils:

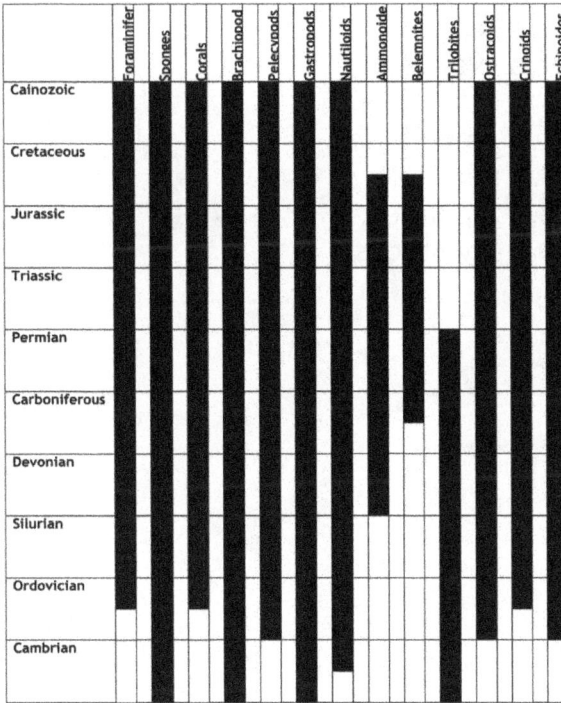

If a rock specimen contains (together) the fossils – *ammonoids, belemnites* and *trilobites*, then the <u>age range</u> for this rock is most likely:

- A. Devonian to Cretaceous;
- B. Jurassic to Devonian;
- C. Devonian to Permian;
- D. Permian to Carboniferous.

10. Vertebrate animals from the fossil record which lay their eggs on land are called:

        A. Amphibates;
        B. Repiloids;
        C. Amniotes;
        D. Viviparous.

Review and Discussion Question

1. Detail the main ways by which ancient organism can be preserved.

2. Why are fossils rare to find? Give some reasons why fossils may not be found in a locality.

3. What are pseudofossils? Give some examples.

4. Why is taxonomic classification necessary? Outline the binomial system of Linnaeus.

5. Distinguish between a pinnae and a pinnule in ferns.

6. What are the major differences and similarities between brachiopods and molluscs?

7. Fossils have been used as indicator tools in geological field work. Explain at least three major uses for indicator fossils in modern geology.

8. Suggest a possible environment for each of the following fossils:

        (a) *Dicroidium*
        (b) *Lovenia*
        (c) *Favosites*
        (d) *Pachypleurosaurs*
        (e) *Eohippus*

9. Why are vertebrate fossils more difficult to find than invertebrate fossils?

10. What were the geological Periods noted for each of the following fossils:

> (a) brachiopods
> (b) fish
> (c) corals
> (d) trilobites
> (e) mammals

11. Outline briefly the steps which would be taken

> (a) before and
> (b) after

mounting a personal hunt for fossils in a selected area.

12. Studies of a particular group of trilobites show a number of significant changes in general body structure over time from oldest to latest due to evolution. These changes include a(n):

- decrease in the number of thoracic segments (not lobes)
- enlargement of the pygidium
- increase in ornamentation of and spines at the end of the pleura
- shortening of the glabella with fewer segments
- an increase in eye size

Given the nature of the ancient form (shown below)

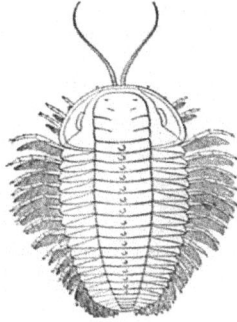

Draw a simple sketch of a possible species which may have later evolved from this species (Note: this is a good exercise with no exact correct answer)

13. A shale in a certain location was found to contain the following plant fossils:

> *Dicroidium dubium,*
> *D. zuberii,*
> *D. elongate*   and
> *Taeniopteris sp.*

Upon returning to base and consulting reference books, the following table was found:

GENUS & SPECIES	GEOLOGICAL PERIODS				
	Late Permian	Triassic			Early Jurassic
		Early	Middle	Late	
Glossopteris species	▬▬				
Voltziopsis sp.	▬▬▬				
Danaeopsis hughesii	▬▬▬▬				
Dicroidium brownii	▬▬▬				
D.zuberi		▬▬▬▬▬▬			
D.dubium		▬▬			
D.narabeenense		▬▬▬			
D.lancifolium		▬▬▬▬▬▬▬▬			
D. elongata			▬▬▬▬▬▬		
Pleuromeia sp.		▬▬			
Pachydermophyllum sp.			▬▬▬		
Taeniopteris sp.			▬▬▬		

Use this table to determine the age of the shale which contained the fossil plants. Give the scientific laws or principles which enabled your conclusion.

14. Distinguish between extinct, extant and endangered when referring to living things. Give an example from each term to support your answer.

15. Research the internet to find out about species of <u>animals</u> which have been declared extinct in the last 50 years

Answers to Multichoice Questions

Q1. C Q2. D. Q3. D Q4. B Q5. C Q6. B Q7. C Q8. A Q9. D Q10. C

Reading List

Alvarez LW, Alvarez W, Asaro F, Michel H.V. (1980). Extraterrestrial cause for the Cretaceous-Tertiary extinction. *Science* 208 (4448): pp.1095-1108. Bibcode:1980Sci..208.1095A.doi:10.1126/science.208.4448.1095.PMID 17783054.

Azzaroli, A. (1998). The genus Equus in North America. *Palaeontographltal* 85: pp. 1-60.

Cavalier-Smith, T. (2007). A revised six-kingdom system of life. *Biological Reviews* 73 (3): pp203-266.

Choi, C.Q. (2013). *Asteroid Impact That Killed the Dinosaurs: New Evidence.* Live Science Contributor. February. http://www.livescience.com/26933-chicxulub-cosmic-impact-dinosaurs.html

J. Cook, et al. (2013) Quantifying the consensus on anthropogenic global warming in the scientific literature. *Environmental Research Letters* Vol. 8 No. 2, (June); DOI:10.1088/1748-9326/8/2/024024.

Cowen, Richard. (1994). Tracking the Course of Evolution – The K-T Extinction. In *History of Life,* 2nd. Edition. Boston, Massachusetts:Blackwell Science 460 pp.

Endangered Species International.(2011).*The Five Worst Mass Extinctions*
http://www.endangeredspeciesinternational.org/overview.ht ml

EPA (the US Environmental Protection Agency) *Causes of Climate Change*. Last updated on 21/7/2015.
http://www.epa.gov/climatechange/science/causes.html

Fortey, R. (1999). *Life: A Natural History of the First Four Billion Years of Life on Earth*. Vintage. pp. 238-260. ISBN 978-0-375-70261-7.

Foster, J. R. and Lucas, S. G. (eds.). (2006). *Paleontology and Geology of the Upper Jurassic Morrison Formation*. New Mexico Museum of Natural History and Science Bulletin **36**. Albuquerque: New Mexico Museum of Natural History and Science (pp. 131-138).

Glaessner, M. F. & Daily, B. (1959). The Geology and Late Precambrian Fauna of the Ediacara Fossil Reserve. *Records of the South Australian Museum* 13: pp.369-407.

Gon III, S. (2015). A Guide to the Orders of Trilobites - A website devoted to understanding trilobites.
http://www.trilobites.info/

Gribbin, J.R. (1982). *Future Weather: Carbon Dioxide, Climate and the Greenhouse Effect*. New York:Penguin.

Imbrie, J. & Imbrie, K.P. (1979). *Ice ages: solving the mystery*. Short Hills NJ: Enslow Publishers. ISBN 978-0-89490-015-0.

Ivanov,M., Hrdlickova, S & Gregorova, R. (2001). *The Complete Encyclopedia of Fossils*. Praha, Solvakia: Rebo. pp.312.

Kolbert, E. (2014). *The Sixth Extinction: An Unnatural History*. New York: Henry Holt and Company. ISBN 978-0805092998.

Kunzig, R & Broecker, W. (2008). *Fixing Climate - The Story of Climate Science and How to Stop Global Warming*.London: Green Profile. 288 pages.

Lambert, D. and the Diagram Group. (1985). *The Field Guide to Prehistoric Life*. New York: Facts on File Publications.

Leakey, R. & Lewin, R. (1996).*The Sixth Extinction: Patterns of Life and the Future of Humankind*. New York: Anchor Books. ISBN 0-385-46809-1.

McLoughlin, S.(2013).The fossil flora of Dinmore. *Australian Dinosaurs* Issue 10 pp 40-49 Febuary.

Moore, R.C., Lalicker, C.G and Fischer, A.G. (2004). *Invertebrate Fossils*. New York: McGraw-Hill. ISBN 10: 8123911394

NewWorld Encyclopedia - Trilobites. (2008). http://www.newworldencyclopedia.org/entry/Trilobite.

Prothero, D. R. and Shubin, N.(1989). The evolution of Oligocene horses. *The Evolution of Perissodactyls*. pp. 142-175. New York: Clarendon Press.

Rey L.V, Holtz Jr T.R. (2007). *Dinosaurs: the most complete, up-to-date encyclopedia for dinosaur lovers of all ages.* New York: Random House.

Royal Society(UK) & the US National Academy of Sciences (US). (2014). Climate Change – evidence and Causes: An overview from the Royal Society and the US National Academy of Sciences, February. ISBN 978-0-08-020409-3. http://dels.nas.edu/resources/static-assets/exec-office-other/climate-change-full.pdf

Scott, P.T. (1984). Some Common Invertebrate & Plant Fossils. . *Science & Ag. Bulletin.* 6 (3) Tamworth NSW, NW Region, NSW Dept. Ed., pp 46-66

Tanner LH, Lucas SG & Chapman MG. (2004). Assessing the record and causes of Late Triassic extinctions. *Earth-Science Reviews* **65** (1-2): pp. 103–139. Bibcode:2004ESRv...65..103T.doi:10.1016/S0012-8252(03)00082-5.

White, M.E. (1988). *Australia's Fossil Plants.* French's Foreset, NSW: Reed Books. 144 pp. ISBN 0 7301 0259 9

Wooldridge, S. A. (2008). Mass extinctions past and present: a unifying hypothesis". *Biogeosciences Discuss* (Copernicus) **5** (3) June: pp. 2401–2423. doi:10.5194/bgd-5-2401-2008.

Yancey, T. E., Garvie, C. L. & Wicksten, M. (2010). The Middle Eocene *Belosaepia ungula* (Cephalopoda: Coleoida) from Texas: Structure, Ontogeny and Function. *Journal of Paleontology* **84** (2): pp. 267-287. doi:10.1666/09-018R

key terms index
(page numbers are in brackets)

adductor scars (57) are the scars left by the adductor muscles on the shells of molluscs which were used to open and close the shells.

albedo (82) is the reflection of radiation from the earth, notably from the ice caps and upper atmospheric clouds giving a generally cooling effect.

amniotes (63) are animals which lay eggs, having a leathery skin or shell, on land.

Anthropocene Epoch (91) a term coined for this latter part of the Holocene Epoch and relating to changes made by humankind.

anthropogenic (87) is any effect which is the result of human activity, especially those which have contributed to global warming, such as deforestation and burning wood and fossil fuels (coal, oil, peat, gas).

articulate (52) in bivalves, is the possession of toothed hinges which open the tests or shells.

assemblage (21) is a group of fossils found together and usually representing a specific age.

bifurcation (33) is the splitting of veins within a fern pinnule into two, usually near the end of the vein.

binomial system (12) is the way living things and fossils are classified, eventually using two names of their genus and species to identify individuals.

bipinnate fronds (32) is when a fern has a stem (or rachis) consisting of two rows of leaflets (pinnae) of many smaller pinnules on both of its sides.

bilateral symmetry (60) is found in some animals where the left side appears identical to the right side e.g. many vertebrates.

bivalves (51) meaning two shells or valves such as in brachiopods and some molluscs.

calyx (47) is the main body of a crinoid at the end of the stem.

calcareous (47) refers to shells etc. being made of calcium carbonate.

camerae (56) are the hollow cavities within ammonites which are connected by a hollow tube called the siphuncle.

carbonization (5) is fossilization of plants by the decomposing of the organic matter to carbon.

cast (5) is a fossil made by the infilling of a mold which then gives the shape of the original organism.

cephalon (58,71) is the head of a trilobite.

cephalopods (55) from the Greek for head foot, these are molluscs with a prominent head, and a set of tentacles.

chromosomes (70) are seen within the nucleus of a cell as it starts to reproduce by cell division. they contain the coded bundles called genes.

cilia (51) are fine hairs at the outer edge of the valves of brachiopods which move to create a water current which enables them to filter food particles out of the water as well as being used for respiration

columella (43) is a raised central section in the centre of a corallite where the septae meet.

comparative anatomy (1) is the study of similarities and differences in the internal and external body structures of different species of living things.

concretions (8) are weathering structures formed in sedimentary rocks which may resemble eggs and other fossils.

cone-in-cone structures (8) are false fossils composed of concentric cones of calcite or gypsum mineral, with thin layers of clay between the cones possibly caused by crystal fibre growth.

coprolites (6) are the solid excreta passed by the ancient organism and then fossilized.

corallites (41) are the hexagonal compartments which contain the soft-bodies polyps of colonial corals.

corallum (41) is the larger body of colonial corals consisting of many linked corallites which hold the individual animal or polyp

correlation (22) is the comparison of rock layers or sequences using mutual specimens of fossils found within them.

cotyledons (38) are the seed leaf or cover in flowering plants (angiosperms) which protect the plant embryo within the seed. in plants such as grasses and grains (monocotyledons) there is only one seed cover but there are two in most other flowering plants (dicotyledons).

cyanobacteria (26) are blue-green algae organisms.

cynodonts (68) meaning dog-teeth, were mammal-like therapsid animals with sturdy bodies with their legs underneath.

cytoplasm (40) is the jelly-like material of some single-celled organisms which are sometimes extruded as pseudopod or false feet.

dendrites (7) are fine cracks in rock which are often filled with mineral and resemble ferns.

dinosaurs (64) are the diverse group of animals which emerged in the Mesozoic Era. the term is often used for any large animal of this era.

dissepiments (50) are the upper and lower parts of the fenestrules or windows of fenestella, connecting the carinas or ridges of pores holding the individual zooids.

DNA (70) or deoxyribose nucleic acid is found in the genes of the cell nuclei of organisms and which determines the characteristics of the individual.

eccentricity (83) are changes to the elliptical shape of the orbit.

encasement (6) is fossilization by trapping the organism within a resistant substance e.g. insects being trapped in tree sap which later hardens to amber.

epitheca (41) is the outer wall of the corallite and made from aragonite.

evolution (69) is the concept that organisms can slowly change into other forms as their genetic code changes slightly due to natural radiation giving a new generation which is more suited to the changing environment.

exoskeleton (58) is the hard outer covering, usually jointed which supports and protects and allows movement in some smaller animals such as the arthropods.

extant (75) meaning that the organism is still existing and not extinct.

extinct (2) is when a species of living things ceases to exist.

facial sutures (58) are the grooves or joints within the head of a trilobite.

fenestrules (49) meaning windows, are rectangular gaps within the colonial body of bryozoans such as *fenestella* species.

foramin (51) is the hole in the pedicle shell of a brachiopod through which the pedicle or foot can be extruded.

fossils (1) named from the Latin *fossilis* meaning to obtain by digging and refers to any remains or traces of ancient life.

fossil horizon (10,28) is a very thin layer within a sedimentary rock which contains fossils.

fulgurites (9) from Latin *fulgur*, meaning lightning, are found at the base of pure silica sand dunes and are the result of lightning strikes on the top of the dune which fuses the silica sand together.

furcula (67) was the wish bone structure in some smaller theropods which assisted in the later flight of birds which evolved from them.

Gaia hypothesis (92) is a questionable idea that the earth as a system is self-healing and will overcome any harmful effects made by humans.

genes (70) are the structures made up of twisted strands of deoxyribose nucleic acid (dna) within the nucleus of all cells which contain the genetic codes for all of the features of an organism. these genes make up the chromosomes seen within the nucleus as the cell reproduces by cell division.

glabella (58) is the raised section of the head (cephalon) of trilobites.

Gondwana (81) or Gondwanaland (named after the Gondwana region or "the land of the Gonds", of northern India), was the ancient southern continent which combined with the ancient northern continent of Laurasia, formed the super-continent Pangea. Gondwana existed from the late Paleozoic to early Mesozoic Eras and eventually broke up to form the modern continents of Antarctica, Australia, South America and Africa as well as the Arabian and Indian sub-continents.

greenhouse effect (79) is a natural process whereby infra-red radiation from the sun passes through the atmosphere and heats the surface which then re-radiates it back as longer wavelengths which are absorbed by gases in the atmosphere which is then heated.

half-life (23) is the time taken for a particular radioactive isotope to decay to one-half of its original radioactivity or mass to a series of other isotopes (daughter products) sub-atomic particles and radiation.

ichiofossils (6) from the Greek, *ikhnos*, for trace or track, are remains of an ancient organism's activities including tracks, burrows, coprolites (solid waste pellets) and eggs, which may be later mineralized or formed as molds and casts.

impressions (5) fossilization by complete removal of the organism leaving a hollow (called a mold) which may then be filled by material which forms a replica of the original (called a cast).

index fossils (26) are an assemblage of fossils which are specific to certain geological times and are used for identification or correlation.

invertebrates (40) are animals which do not have a bony or cartilaginous backbone.

iridium (77) is one of the 92 naturally-occurring elements (types of atoms), symbol Ir and atomic number 77 belonging to the transition metals, silvery white in colour and dense. it is very rare on earth but is more abundant in meteorites (meteors which strike earth).

isotopes (80) are different forms of the one chemical element. each of the 92 natural elements (and those artificially made) have isotopes due to the different numbers of neutrons in their nucleus e.g. normal carbon has the isotopes of carbon 12 with 6 protons and 6 neutrons in its nucleus, but carbon 14 has two extra neutrons and so is radioactive.

Law of Superposition (21) is a fundamental law of sedimentation proposed by Nicholas Steno in the 17th century that the oldest beds are deposited below younger beds.

leaf scars (29) are diamond-shaped scars seen on the trunks of some palms and fern-trees where the fronds or branches were attached and have now fallen off.

lobate (35) is the shape of fern pinnules which are broad along the rachis and have a rounded apex or end.

lophophore (51) is feeding ring of cilia (hair-like tentacles) surrounding the mouth of bryozoan and brachiopods.

mantle (53) is a cape of tissue as flaps protruding from molluscs which are used for respiration, digestion, excretion and sometimes movement. in some species it secretes calcium carbonate and forms a shell.

midrib (33) is the central supporting vein of a fern pinnule.

mould (5) is a hollow space or impression left by an organism.

mural pores (41) are holes within the walls of the corallites to allow the flow of water and nutrients.

mutations (69) are permanent changes in organisms due to changes in their DNA which determines the organism's characteristics.

nothosaur (caption 65) was a semi-aquatic marine reptile with long necks and webbed feet.

notochord (61) is a primitive spinal column and is only found in the early members of phylum chordata.

nuclear membrane (13) is the thin skin surrounding and holding the contents of cell nuclei, the spherical structures within cells which function. absent in prokaryotic organisms.

obliquity (83) is the tilt of the earth's axis which is at 23.44 degrees.

palaeontology (2) is the scientific study of the remains and traces of once living things.

pallial line (57) is where the mantle attached to the valves (shells) in bivalve molluscs.

pedicle (51) is the muscular foot which emerges from an opening (foramen) in one of the shells (the brachial valve) of a brachiopod.

pentadactyl limb (72) is the limb (arm, leg, flipper, wing) of mammals which consists of one upper bone connect at one end to the shoulder and to two bones in the lower limb. these are then connected to a series of smaller bones forming a five-digit hand.

periods (22) referring to geological time, are the main working units of time used by palaeontologists. initially given names of ancient tribes, localities or descriptive terms, they were later delineated by absolute ages in millions of years ago.

permafrost (85) or frozen soil (which contained water) existing in the high latitudes, notably in Siberia, Greenland and Canada.

permineralisation (3) is fossilization by minerals from solution which fills the spaces within the organism.

photosynthesis (26) is the biochemical process in all organisms containing the green pigment chlorophyll which is used to make simple sugars for food using sunlight, carbon dioxide gas and water and liberating waste oxygen.

phragmocone (55) is the long, internal cone-like structure of belemnites, squids and similar gastropods, ending in a hard protective structure called a guard.

pinnae (29) are the simple structures in ferns as a frond, resembling a collection of leaves.

pinnules (29) are the smaller, individual leaf-like structures on the fronds or pinnae of ferns.

placoderms (61) meaning platy skinned were armoured fish having a hard external covering, especially around the head.

planispiral (56) is when a shell is coiled in a single horizontal plane (i.e. flat) and the diameter increases away from the axis of coiling.

polyps (41) are the individual invertebrate animals of colonial organisms such as corals.

pygidium (35) is the tail of a trilobite.

precession (83) is the way that the earth wobbles around the axis through its poles.

pseudofossils (7) meaning false fossils and are structures formed by processes other than fossilization of life e.g. dendrites which are fine, branch-like cracks filled with mineral.

pseudopods (40) from the Latin for false feet, are projections of jelly-like cytoplasm extruded from the tests of some invertebrates used to obtain food.

pterosaurs (65) meaning winged lizard, were flying reptiles which existed from the late Triassic to early Cretaceous Periods. they had hollow bones and a thin membrane stretched between their ankles and an extended fourth finger.

pygidium (58) is the tail at the end of the segmented body of trilobites.

rachis (29) is a simple support column in ferns and other similar plants resembling the modern stem in modern plants but without their complex structure.

radial symmetry (48) bodies with five identical sections radiating out from a centre e.g. starfish and other echinoderms.

radula (53) is the rasp-like tongue of molluscs.

recrystallization (4) is fossilization by the chemical conversion of the mineral of the organism by a more stable form of mineral e.g. aragonite to calcite in corals.

replacement (4) is fossilization by complete filling by a resistant mineral of the space left as the entire fossil disintegrates.

reticulate venation (30) is the pattern seen in the veins of leaves and ferns where the vein branches out many times but later re-joins.

rhabdosomes (45) is the name given to the whole colonial structure of a graptolite.

rostrum (55) is a heavy calcite guard surrounding the long chambered phragmocone or chambered end-section of belemnites.

sauropods (65) were dinosaurs having large bodies with trunk-like legs and long necks e.g. brontosaurus.

sedimentary rocks (1) rocks formed from sediments such as silt, sand and gravel carried down by water, wind or ice and then settled.

septae (43) are internal partitions within each corallite or individual coral structure.

sicula (45) is the end of the stipe of a graptolite colony.

siphuncle (56) is a hollow tube which connects the chambers of ammonites which allowed the removal of water and also assisted in changing the buoyancy of the animal.

six kingdom system (13) is a system of biological classification using six distinct major groups or kingdoms, of living things. it is based on that devised by Thomas Cavalier-Smith in 2007.

spathulate (30) are leaves beginning from a tapered part near the stem but becoming broader at the end.

species (19) is a group of organisms within which the individuals may mate and give rise to new offspring.

spores (29) is a one-cell reproductive unit found in simple plants such as ferns. often found in special capsules or on the under surface of pinnae (leaflets) and are distributed mostly by water.

stipe (45) is the long, thread-like structure of graptolites which hold the individual cups or theca containing the animals.

stolon (45) is the hollow tube which runs down inside the long thread-like structure or stipes of graptolites.

stromatalites (26) from Greek *strōma* for mattress and lithos for rock, are layered bio-chemical structures which build up layer upon layer in shallow water by the trapping, binding and cementation of sedimentary grains by microorganisms, especially cyanobacteria.

tabulae (41) are the internal partitions within the hollow structures or corallites of corals and similar animals acting like a floor in a modern high-rise.

tabulate (43) are the flat-topped colonial corals.

taxa (12) plural of taxon and is a group of one or more populations of an organism or organisms which form a unit.

taxonomy (12) is the classification of organisms which can also be applied to fossils.

tektites (77) are small, dark-coloured glass formed by the melting and fusion of rock and soil from the impact of a meteorite.

tests (39) are the shells of invertebrate animals.

theca (45) is the small cup-like structures which contain the soft-bodied individuals of colonial graptolites.

therapsids (68) are animals with legs under their bodies and with teeth and skull features similar to modern mammals.

theropods (66) were smaller dinosaurs having hollow bones, bipedal gate (walked on two legs) and a rib cage with a furcula or wishbone.

thorax (58) is the central body of a trilobite.

tipping point (80) is the time when the change in global climate may become irreversible.

type locality (12) is the precise location where a new fossil has been discovered and is written in the literature for further reference.

unconformity (22) is a stratigraphical term (i.e. to do with layers of rock) representing a time of erosion and then further sedimentation. this often causes a gap in the fossil record.

valves (51) are the shells in some molluscs and brachiopods.

vascular tissue (29) consists of specialized tubes within plants which carry water, food and air in more advanced plants such as angiosperms.

vertebrates (17) are animals which have an internal skeleton with a nerve cord with a spinal column of vertebrae, rings of bone linked together.

whorls (33) are an arrangement of petals, leaves or branches that radiate from a single point and surround or wrap around the stem.

zooids (45) are the living polyps of a graptolite colony or rhabdosome.